Richard Parkinson
Princeton 1978

AERODYNAMICS

Dr. Theodore von Kármán has taught in several colleges and universities. Formerly Director of the Guggenheim Aeronautics Laboratory, and Chairman of the Scientific Advisory Board of the USAF, he was also Chairman of the Advisory Group for Aeronautical Research and Development.

Cornell University Press

Aerodynamics

Theodore von Kármán

McGraw-Hill Book Company, Inc.
New York Toronto London

Cornell University Press

London: Oxford University Press

First McGraw-Hill Paperback Edition, 1963

78910 MUMU 7

ISBN 07-067602-x

PRINTED IN THE UNITED STATES OF AMERICA

To the memory of my sister

JOSEPHINE DE KÁRMÁN

whose devoted companionship

secured for me the peace of mind

necessary for scientific thinking

» *Preface*

THIS brief account of the main principles of the science of aerodynamics and the historical development of aerodynamical thinking was planned to appear in 1953, the anniversary year of powered flight. The pressure of business prevented me from finishing the manuscript on the date originally planned, but I am now happy that this little book will come to the reader after the flood of publications devoted to the magnificent accomplishments of the first half-century of the air age has subsided.

In this volume my purpose is not to present a sentimental or emotional review of the achievements of aviation in the past fifty years but rather, first, to give some idea of aerodynamic thought to readers familiar with the facts of aviation but less familiar with the underlying theories; second, to remind persons engaged in the study or professional use of aerodynamic science how much mental effort was necessary to arrive at an understanding of the fundamental phenomena, which the present-day student obtains readily from books and lectures.

I want to express my sincere appreciation for the help and assistance that I received from many persons. First of all, both in the preparation of the lectures that led to the publication of this book and in the preparation of the final manuscript, I was splendidly supported by William R. Sears, Mabel R. Sears, and many members of the Graduate School of Aeronautical Engineering and of the Faculty of Cornell University. Professor Itiro

Tani contributed many data and references and made special studies on some of the questions touched on in the book. I am grateful to Dr. Frank J. Malina for reading the manuscript. Many of his valuable suggestions have been included. Finally, I am indebted to the Cornell University Press for editorial assistance and skillful technical production.

THEODORE VON KÁRMÁN

Pasadena, California
March 1954

» Contents

Aerodynamics: SELECTED TOPICS IN THE LIGHT OF THEIR HISTORICAL DEVELOPMENT

Leonardo da Vinci (self-portrait)

Sir Isaac Newton

Sir George Cayley

Jean Le Rond d'Alembert

CHAPTER I

» *Aerodynamic Research before the Era of Flight*

IN 1953 we celebrated the golden anniversary of human flight. The development of the flying machine from the rather primitive contraption of the Wright brothers to the complex and efficient high-speed airplane of today has been most spectacular. Yet when I fly in bumpy weather or when I am forced to wait hours at an airport because of the weather—or because of the ignorance of the weatherman—I wonder whether our achievement is really so miraculous. We have, nevertheless, succeeded in passing through the sonic barrier, and the records for endurance and nonstop distance are far beyond the figures I would have thought possible forty-five years ago when I first became interested in aeronautical research.

When, however, I recall the state of knowledge at that time concerning the mechanics of flight and the theory of airflow, it appears to me that, parallel to the development of the art of aeronautical engineering, the science of aerodynamics has traversed a path scarcely less significant. Our knowledge of the reasons "why we can fly" and "how we fly" has increased both in scope and depth in a rather impressive way.

A short summary of the fundamental aspects of this scientific progress is the subject of this volume. Many books have been published on the history of aviation—the history of the conquest of the air. In this book, however, I am not concerned with the progress made in aircraft structures—or, more generally, in aircraft design. Instead I want to report on the progress made in aerodynamics, which is one of the branches of theoretical physics. My subject is not as spectacular as some other branches of theoretical physics which have become extremely popular for several reasons.

Some branches of theoretical physics lend themselves to speculation on the origin and true nature of the universe, others to questioning of philosophical beliefs, such as the laws of causality, commonly accepted for centuries. Finally, fundamental progress in physics has led to technical applications of horrifying nature and energy production of unheard-of magnitudes. The reader will guess that I have in mind, in particular, the theory of relativity, quantum mechanics, and nuclear physics. We aerodynamicists were always more modest and did not attempt to change basic beliefs of the human mind or to interfere with the business of the good Lord or divine Providence!

Nevertheless, I believe that the development of aerodynamic science during this half-century of human flight should be of general interest beyond the limits of aeronautical circles. It is a rare example of co-operation between "men of mathematics"—as my friend Eric T. Bell calls them—and creative engineers. Mathematical theories from the happy hunting grounds of pure mathematicians were found suitable to describe the airflow produced by aircraft with such excellent accuracy that they could be applied directly to airplane design. This is a remarkable fact if we compare it with an opinion expressed by an expert in 1879, which I found in the fourteenth annual report of the Aeronautical Society of Great Britain, the predecessor of the present Royal Aeronautical Society: "Mathematics up to the present day have been quite useless to us in regard to flying."

Credit must be given to the builders of aerodynamic theory in the last half-century for the fact that the foregoing statement is now untrue, and what is much more, even engineers and aircraft designers admit that it is no longer true.

Period of Legend and Artistic Imagination

Every historian of aviation starts with legendary examples, which at least show mankind's yearning to fly like the birds. Most of these well-known stories, however, do not contain many elements of aerodynamic thinking or experience. We have, for example, the myth of Daedalus and Icarus. The only technological factor here is that the fliers did not know about heat-resistant materials; the aerodynamic aspect of the flight is not discussed. In the Bible (Proverbs 30: 18, 19), Solomon, son of David and King of Israel, quotes the words of Agur, the son of Jakeh, saying:

> There be three things which are too wonderful for me,
> Yea, four which I know not.

As the first he lists:

> The way of an eagle in the air.

Here man at least admitted his ignorance of aerodynamics. (The other three "things" I shall let the reader look up. The last is frequently quoted.)

A story told in one of the Eddas, of Norse mythology, shows certain observations of an aerodynamic nature. It seems that one Wayland, a smith whose trade was manufacturing weapons, also constructed wings to be attached to his body. He apparently was a very vicious fellow; since, as shown in the drawing (Fig. 1), he took his enemies aloft and let them drop from a height in order to kill them.

Fig. 1. Wayland the Smith (*above*) in his "feather-dress." (From K. M. Buck, *The Wayland-Dietrich Saga* [London, 1924].)

Now, according to the Saga, written perhaps in the thirteenth century (Ref. 1) but originating in the fifth century or earlier, Wayland after finishing his first set of wings planned with his brother Egil to try them out, i.e., to make a test flight.

His brother asked him, "How shall I do this? I have no knowledge in this field."

> Quoth Wayland, slowly and with emphasis,
>
> . . . . .
>
> "Against the wind shalt thou rise easily,
>
> . . . . .
>
> Then, when thou wouldst descend, fly with the wind."
> Egil put on as told the feather-dress,
> And soon flew high in air swift as a bird,
> Lightly and easily both high and low.
> But when he would alight upon the ground,
> Turning, he flew full quickly with the wind,
> Was headlong borne to earth, and in his fall
> Had much ado to save his neck from harm.

This is what is said in the Eddas.

Then Egil asked Wayland, "How is this? Your wings are good for take-off but not good for landing! I must confess," he added, "that if they were really good, I would have kept them."

Wayland answered:

> "When I bade
> That thou shouldst with the wind make thy descent
> I told thee wrong. . . . I did not trust thee quite.
>
> . . . . .
>
> Remember this, that every bird that flies
> Rises against the wind and so alights."

If we proceed from legend to history, we find that many great men with artistic imagination studied the fundamentals of bird flight and speculated on the possibilities of human flight. The drawings and notes of Leonardo da Vinci (1452–1519) represent an excellent example of such studies (Ref. 2).

It seems that he considered two methods of flight. One consisted of imitation of bird flight. In Fig. 2 we see a man equipped with a pair of wings, beating them like a bird. Today we call an aircraft of this type an *ornithopter*. The other method was based

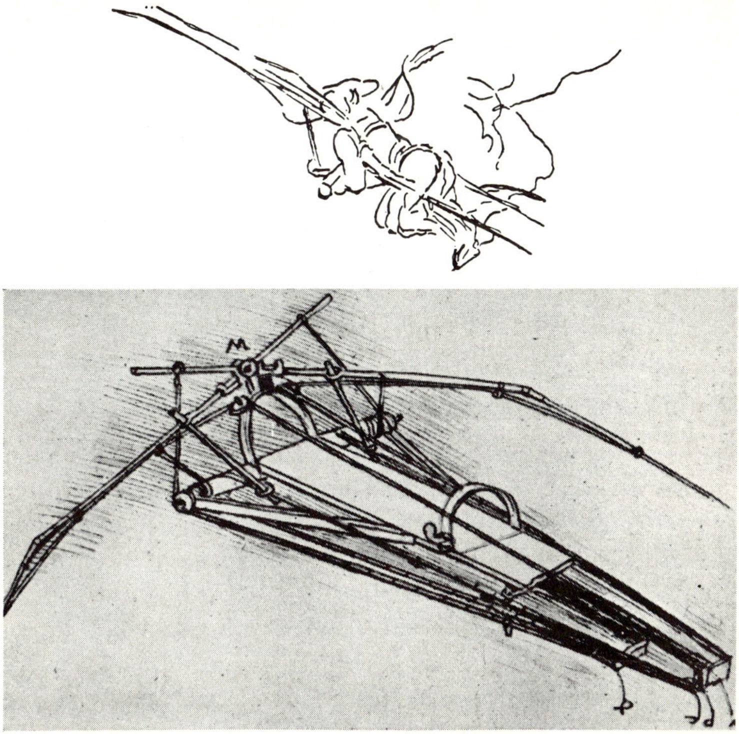

Fig. 2. Leonardo da Vinci's design of an ornithopter.

on a screw—we call it the screw of Archimedes—which would penetrate the air (Fig. 3). This is the predecessor of the present-day *helicopter*. The characteristic feature underlying both proposed systems was the general belief that sustentation of the weight of the body and propulsion should be accomplished by the same mechanism. This is true for the bird, whose propulsion and sustentation are produced by the motion of the same wings. It is also correct in the case of the helicopter. The idea of imitating bird flight was predominant for centuries in the minds of inventors. Some, however, recognized the limitations of mere imitation of nature. As Hiram Maxim, one of the British pioneers in aeronautics once remarked, "The successful locomotive was not based upon an imitation of an elephant."

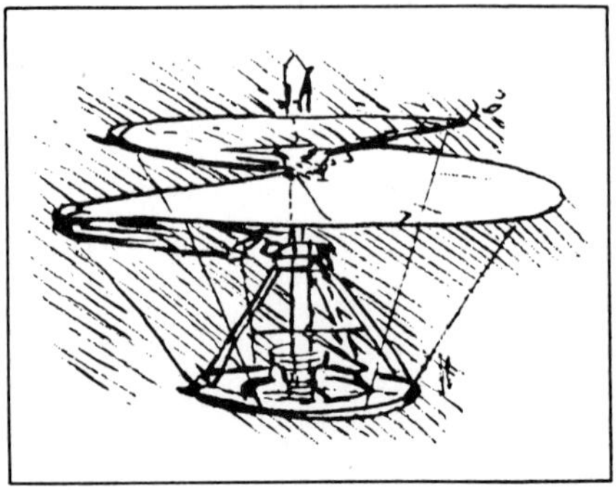

Fig. 3. Leonardo da Vinci's design of a helicopter.

Fundamental Notions: Newton's Law of Air Resistance

I want to restrict myself to dynamic flight, i.e., to aircraft heavier than air. The development of lighter-than-air craft happened more or less independently, at least as far as the free balloon is concerned. The principle of sustentation by hydro- or aerostatic lift has been understood ever since Archimedes stated his famous principle. The successful experiments of the Montgolfiers preceded any serious experiments aimed toward dynamic flight, which means sustentation by forces produced in the air by the motion of solids. Aerodynamics, in addition to aerostatics, entered the problem of balloon flight when propulsion of the balloon was proposed. Such proposals were made very soon after the first successes of free balloons; Benjamin Franklin was one of the first men who did some thinking in this direction, i.e., in the direction of dirigibles.

Let us return to the problem of heavier-than-air craft. As I mentioned above, the concept of sustentation by flapping wings or by a screw preceded that of the rigid airplane.

The idea that sustentation can be accomplished by moving inclined surfaces in the flight direction, provided we have mechanical power to compensate for the air resistance that hinders this motion, was probably clearly defined for the first time by an Englishman, Sir George Cayley (1773–1857)—not Arthur Cayley, the mathematician—in his papers, published in 1809–1810, on aerial navigation (Ref. 3). He belonged to a group of enthusiasts who tried to solve the problem of flight empirically by building models and studying bird flight. However, in his paper he clearly defined and separated the problem of sustentation, or in modern scientific language the problem of lift, from the problem of drag, i.e., the component of total resistance that works against the flight direction and has to be compensated by propulsion in order to maintain level flight.

Cayley made certain statements that show his keen observation of the influence of streamlining on drag—for example, in the case of spindle-shaped bodies. He said, according to a note appearing in his *Aeronautical and Miscellaneous Note-Book* (Ref. 4), "It has been found by experiment that the shape of the hinder part of the spindle is of as much importance as that of the front in diminishing resistance." Cayley was rather skeptical that theoretical science would make important contributions in the field of flight research:

> I fear, however, that the whole of this subject is of so dark a nature as to be more usefully investigated by experiment than by reasoning [by this he obviously meant theoretical reasoning] and in absence of any conclusive evidence from either, the only way that presents itself is to copy nature; accordingly I shall instance the spindles of the trout and woodcock.

In the *Note-Book*, which was published after Cayley's death, we find the drawing reproduced in Fig. 4. Cayley obtained the profile shown in the drawing by measuring the girths of various cross sections of a trout and dividing the measured lengths by three. It is interesting to learn that the shape of his profile almost exactly coincides with certain modern low-drag airfoil sections, as can be seen in the figure.

Thus the principle of the airplane as we know it now, that of the rigid airplane, was first announced by Cayley. But in order to understand the further development of the airplane and to appreciate the difficulties encountered by the pioneers of aviation, we must look at the state of knowledge, in Cayley's time, of aerodynamics or more specifically of the forces exerted on solid bodies moving through a fluid like air. In order to sketch the knowledge and opinions prevailing at that time, we have to go back to the era in which the science of mechanics was founded.

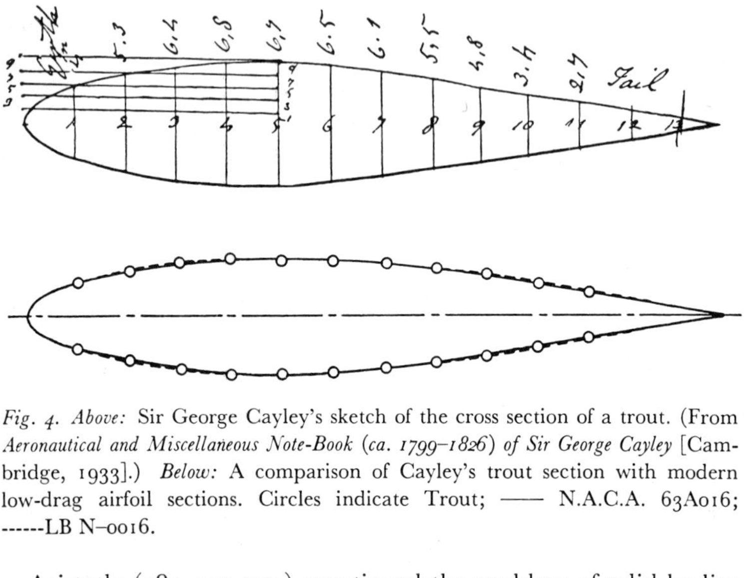

Fig. 4. Above: Sir George Cayley's sketch of the cross section of a trout. (From *Aeronautical and Miscellaneous Note-Book (ca. 1799–1826) of Sir George Cayley* [Cambridge, 1933].) *Below:* A comparison of Cayley's trout section with modern low-drag airfoil sections. Circles indicate Trout; —— N.A.C.A. 63A016; ------LB N–0016.

Aristotle (384–322 B.C.) mentioned the problem of solid bodies moving in air. But, since he believed there is always a force necessary to sustain a uniform or even a decelerated motion, he looked for a force which pushes forward a flying ball, instead of looking for a force which resists the motion.

Galileo Galilei (1564–1642) recognized the law of inertia and had a correct notion of air resistance. He observed that the movement of a pendulum was slowly amortized by air resistance and actually tried to determine the dependence of air resistance on velocity.

The first theory of air resistance deduced from the principles of mechanics was, however, given in the *Philosophiae Naturalis Principia Mathematica* of Sir Isaac Newton (1642–1727) (Ref. 5). First, he clearly stated that the forces acting between a solid body and a fluid are the same whether the body moves with a certain uniform velocity through the fluid originally at rest or the fluid moves with the same velocity against the body.[1] Then in the thirty-third proposition of Section 7 of Book II he made three general statements, valid for bodies of similar shape. These three statements say that the forces acting on two geometrically similar bodies which move in fluids with different densities are proportional to—

a) the square of the velocity,
b) the square of the linear dimension of the body, and
c) the density of the fluid.

These statements follow, according to Newton, from the fundamental laws of mechanics by the following argument: Consider the body at rest, exposed to an originally uniform fluid stream of given velocity. The force acting on the body may be caused by centrifugal forces due to deflection of the fluid or by impact of the fluid particles. In both cases the rate of change of momentum (momentum = mass × velocity) produced in the fluid is proportional to the density of the fluid and the square of the velocities of the individual particles involved in the motion, therefore, supposing similarity of the flow, proportional to the square of the undisturbed stream flow velocity.

Since, according to Newton's general laws, the force acting on a body or particle is equal to the change of its momentum, all forces produced in the fluid and also the resultant force acting between the solid body and the fluid are proportional to the

[1] In Newton's mechanics, this statement appears as a special case of his principle of relativity. As to the interaction between a body and an airstream, it was announced by Leonardo, who said, "The resistance of an object against air at rest is equal to the resistance of the air moving against the object at rest" (Ref. 6).

density of the fluid and the square of the velocity of the fluid stream. The proportionality of the force with the square of the linear dimensions of the body follows easily from consideration of the geometrical similarity, since only pressure forces are considered.

The formula or law known generally as Newton's sine-square law of air resistance refers to the force acting on an inclined flat plate exposed to a uniform airstream. It was much discussed in connection with the problem of flight; in fact it cannot be found in Newton's works. It was deduced by other investigators based on a method of calculation which Newton used for comparison of the air resistance of bodies of different geometrical shapes. In the thirty-fourth proposition of his book he calculated the total force acting on the surface of spheres and cylindrical and conical bodies by computing and adding the forces caused by the impact of air particles, which supposedly move in a straight line until they hit the surface. The same idea applied to the calculation of the force acting on an inclined flat plate leads to the formula

$$F = \rho S U^2 \sin^2\alpha \ ,$$

where ρ is the density of the fluid, S the area of the plate, U the velocity of the plate, and α the angle of inclination. The force F is directed normal to the plate. The quantity $\rho S U \sin\alpha$ is evidently the mass flow in unit time through a cross section, $S \sin\alpha$, equal to the projection of the plate perpendicular to the original flow direction (Fig. 5). It is supposed that after the impact the particles follow the direction of the plate. Then one obtains the change of the momentum of the fluid mass

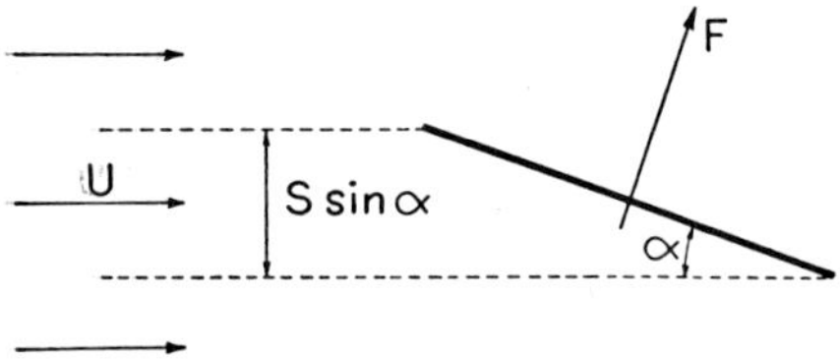

Fig. 5. Diagram illustrating Newton's theory. The mass of fluid deflected by the plate is assumed to be that flowing through the cross section $S \sin\alpha$. U is the velocity of flight, S the area of the plate, α the angle of inclination, and F the force.

hitting the plate in unit time by multiplying this mass by the velocity component, $U \sin \alpha$, created by the impact.

We note that only the dependence of the force on the angle of inclination was computed according to a particular assumption concerning the nature of fluid flow, whereas its dependence on density, dimensions, and velocity was determined by general mechanical principles.

Experimental Techniques in Early Aerodynamics

In the two centuries between the publication of Newton's *Principia* and the date of the first mechanical flight, a great number of observations were made to determine the resistance experienced by a body. Newton's argument had one great merit. He talked in general about *fluids* and pointed out that the same law applies to water as well as to air. The forces are proportional to the respective densities. This statement made it possible to apply the result of measurements made in water to motion in the air, and vice versa. This, of course, represented great progress.

In the long list of experimenters, engineers, and physicists, we find the names of many generally known scientists. Edme Mariotte (1620–1684) measured the force acting on a flat plate submerged in a stream of water. Jean Charles de Borda's (1733–1799) experiments included bodies of various shapes; he put the bodies in motion in the water by means of a rotating arm, the so-called whirling arm. This method had been used before by Benjamin Robins (1707–1751), who carried out his experiment in air (Fig. 6). The whirling-arm technique has been used up to modern times. It has the disadvantage, however, that after some time the air or water begins to rotate with the arm, and it is difficult to determine exactly the speed of the model relative to the air or water that surrounds it.

Several experimental methods were used for pulling the body whose resistance is to be determined in a rectilinear motion. Jean Le Rond d'Alembert (1717–1783), Antoine Condorcet (1743–1794), and Charles Bossut (1730–1814) towed ship models in still water. This was perhaps the first use of the so-called towing-tank technique. In order to carry models through the air in rectilinear motion, locomotives and, later, motor cars have been used. This method, however, is not very exact. First, it can be used only when there is no wind and, second, it is hard to calculate the influence of the ground.

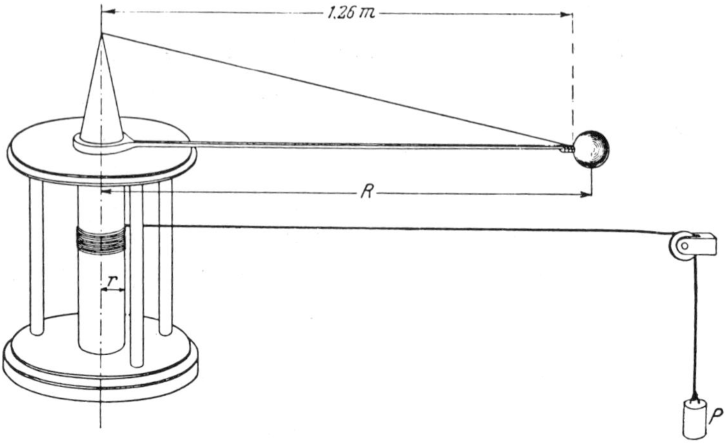

Fig. 6. Whirling arm of Benjamin Robins. (From *Handbuch der Experimentalphysik* [Leipzig, 1931], IV, Teil 2, by permission of Akademische Verlagsgesellschaft.)

Another method of creating a rectilinear motion is to allow a body to fall through the air. Newton himself observed spheres dropped from the dome of St. Paul's Cathedral. Many investigators have used this method. Remarkable experiments were carried out at the end of the nineteenth century and the beginning of the twentieth century by Alexandre Gustave Eiffel (1832–1923) and his collaborators, who used the tower named after Eiffel in Paris (Ref. 7). Fig. 7 shows Eiffel's experimental arrangement for measuring the resistance of a flat plate. The recording instrument, *R*, contained a cylinder that was turned at a rate proportional to the velocity of the falling system relative to the guiding cable. On this cylinder two records were registered. A tuning fork recorded the time. A spring dynamometer inserted between the plate and the supporting frame registered the force acting between these two parts. Now, since the time was given as a function of the displacement, one could calculate the actual acceleration. The difference between the actual acceleration and the acceleration due to gravity is equal to the difference between the force measured and the force due to air resistance divided by the mass rigidly connected with the plate.

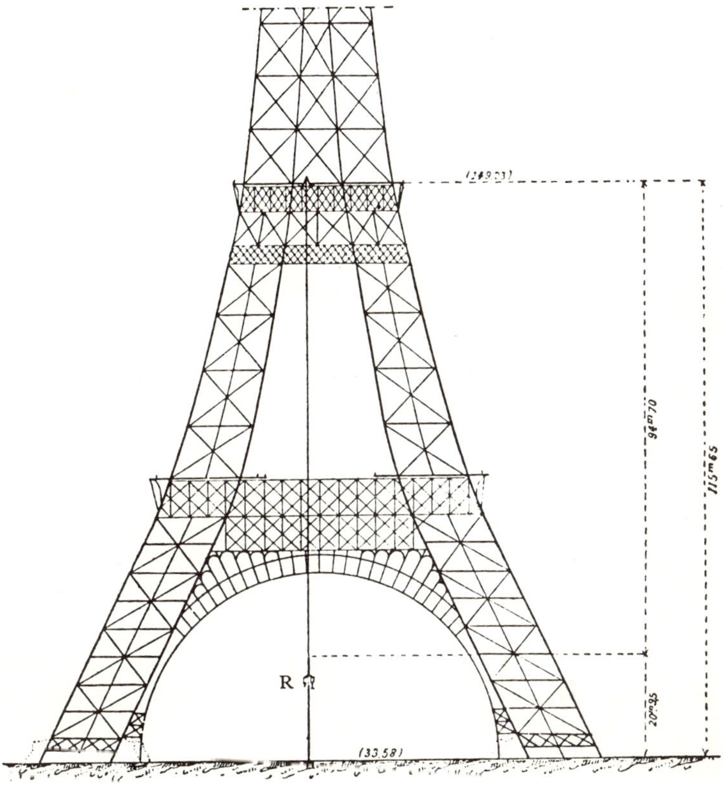

Fig. 7. Eiffel's arrangement for measuring the resistance of a flat plate. (From G. Eiffel, *Recherches expérimentales sur la résistance de l'air exécutées à la tour Eiffel* [Paris, 1910].)

The best method for measuring air resistance is to put a model in an artificial stream of air, i.e., the method of the wind tunnel

(Fig. 8). The first man to make such an installation was Francis Herbert Wenham (1824–1908), founder member of the Aeronautical Society of Great Britain, who designed a wind tunnel in 1871 for that Society. In 1884 another Englishman, Horatio Phillips (1845–1912) built an improved wind tunnel. Following

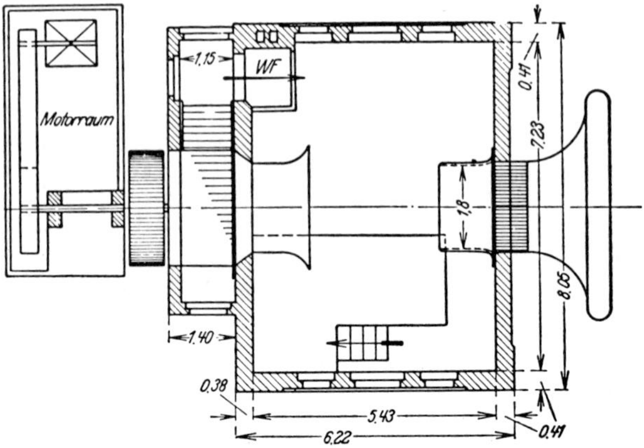

Fig. 8. Old wind tunnel of the University of Aachen; constructed in 1912–1913. The dimensions are given in meters.

these, several other small wind tunnels were built; for example, in 1891 Nikolai E. Joukowski (1847–1921) at the University of Moscow built a tunnel two feet in diameter. In the first decade of our century, wind tunnels were built in almost all countries. Some of the builders were Stanton and Maxim in England, Rateau and Eiffel in France, Prandtl in Germany, Crocco in Italy, and Joukowski and Riabouchinski in Russia. In comparison to the present huge tunnels, these installations were relatively modest. For example, no wind tunnel built before 1910 had more than 100 horsepower. Today a wind tunnel in the French Alps uses hydraulic power to the amount of 120,000 horsepower, and I think the largest wind tunnel that is planned in this country will use about a quarter of a million horsepower for driving the airstream. One of our most powerful tunnels is shown in Fig. 9.

Scientific research, at least as far as means of research are concerned, has followed in general the same scale of development as aircraft construction.

Newton Blamed for Delay in Development of Flight

Experimental evidence has shown that three of Newton's statements were correct: the proportionality to the density, the proportionality to the square of the linear dimension, and the proportionality to the square of the velocity. Of course the last of these is restricted to small or moderate speeds, because, as was known from ballistic experiments even in Newton's time, it does not apply to velocities comparable or superior to the velocity of sound. It applies only so long as the air can be considered as incompressible or of very small compressibility. We shall discuss this question in Chapter IV.

Newton's prediction of proportionality between the force acting on a surface element and the square of the sine of its angle of inclination turned out to be completely false. Experiments show that the force is, rather, nearly linear with the sine of the angle—or with the angle itself in the case of small angles. The question of whether the experimental or Newton's theoretical law is correct has far-reaching consequences in the theory of flight. In fact, if the normal force follows Newton's law, the component forces perpendicular and parallel to flight velocity, i.e., the lift and the drag, are proportional to $\sin^2\alpha \cos\alpha$ and $\sin^3\alpha$, respectively. Thus the lift coefficient, being proportional to the second power of $\sin\alpha$, is very small for small values of the angle α, and if the airplane designer does not want to use large values of α, he needs a tremendous wing area in order to obtain a sufficient amount of lift. On the other hand, the ratio between lift and drag is equal to $\cot\alpha$, and this expression can have a large value only if the angle α is very small. If Newton's law is correct, the poor designer has only the choice of either making a tremendous contraption having a very large

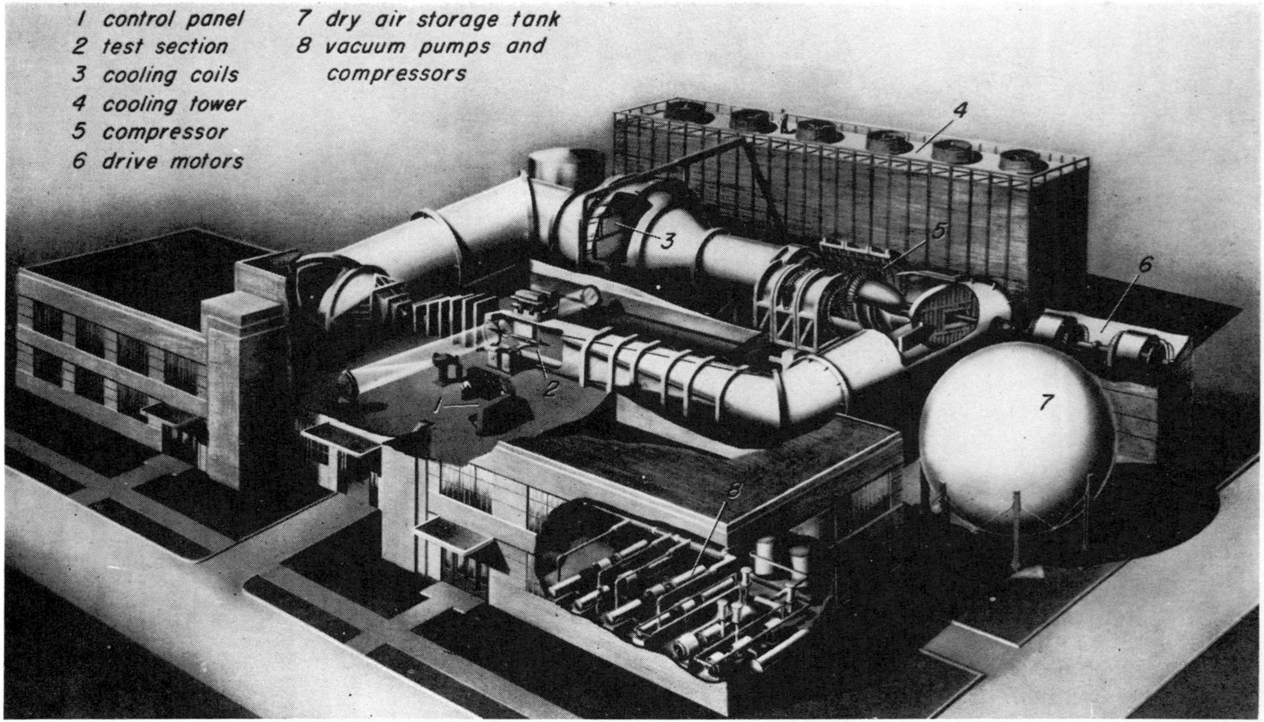

Fig. 9. A modern wind tunnel. Cutaway drawing of the 6-by-6-foot supersonic wind tunnel of the N.A.C.A., Ames Aeronautical Laboratory, Moffett Field, Calif. (Courtesy of the National Advisory Committee for Aeronautics.)

wing area, and therefore a heavy structural weight, or building a machine with a reasonable wing area but low lift-drag ratio, which means a heavy engine for propulsion.

A number of authors expressed the opinion that Newton's law contributed to the pessimistic forecasts one could find in the scientific literature on the possibilities of powered flight. Personally I do not believe that Newton's influence was really so catastrophic. I think most of the people who, in the early period we are talking of, were really interested in flying, did not believe in any theory. But it must be considered that the theory was at variance with the facts. Further it should be noted, as I stated before, that Newton essentially considered blunt or pointed bodies exposed to a parallel stream in order to compare their respective resistances and did not study forces acting on inclined surfaces. We will see later why his theory applied to wing surfaces gave results so different from reality, and on the other hand how his law found new application in the domain of very high supersonic speeds.

Bird Flight: Semiempirical Flight Theories

Throughout the nineteenth century, we observe two practically unrelated developments. On one side, the flight enthusiasts, mostly practical men, developed their own rather primitive theories of bird flight and tried to apply their results to the requirements of human flight. On the other hand, a mathematical theory of fluid dynamics was developed by scientific people; this development was not related to the problem of flight and did not give much useful advice to those who wanted to fly.

The investigations directed toward the realization of human desire to fly dealt especially with two problems: first, to determine the power required for flight; second, to find out the most efficient shapes for wings. Let us consider briefly both problems and the prevailing opinions in this period.

Concerning the question of the power required for flight, the fact that birds actually fly through the air furnished a certain solid support for the speculations. It was recognized at an early date that two characteristic quantities must play an important role in the calculations. One is the ratio between the weight, W, and the wing area, S. We call this ratio, W/S, the wing loading. The second quantity is the ratio between the weight, W, and the power available, P. The ratio, W/P, is called the power loading. In the case of bird flight, the power available is the muscle power that the bird can exert in flight. One may assume that the latter quantity is roughly proportional to the weight of the bird.

Then the main question was to estimate the power required and compare it with the power available. The power required was calculated on the assumption that the soaring bird without working its wings would lose a certain altitude in unit time; this is called the sinking velocity. In order to fly horizontally, the bird has to put in at least as much work as is necessary to raise its body at a rate sufficient to counteract the sinking velocity. This estimate led to the conclusion that the power required per unit weight (i.e., the reciprocal of the power loading) is proportional to the square root of the wing loading.

The general form of this rule was confirmed by a more detailed analysis by Charles Renard (1847–1905) (Ref. 8), one of the leaders in early French aeronautics. He expressed the power required for level flight as the sum of the power necessary for sustentation and the power needed for propulsion, i.e., the drag of the aircraft multiplied by the speed. His formula is quite similar in form to those used in modern aircraft design. Then he computed the velocity at which the power required has a minimum value and substituted this value in his formula. The result is

$$\frac{P}{W} = \text{constant} \times \sqrt{\frac{W}{\rho S}},$$

which corresponds to the expressions obtained earlier for the minimum power necessary for level flight (ρ denotes the density of the air).

The constant in Renard's formula depends on the assumptions made (*a*) for the law of sustentation, and (*b*) for the drag coefficient of the aircraft. The first assumption is the important one.

If Newton's resistance law is used for the computation of the force of sustentation, as we indicated above, a terrific figure is obtained for the necessary power. The result of the computation is more reasonable if the lift is computed by one of the empirical formulas found by experiment. According to Henry, a contemporary of Renard (Ref. 9), the constant in the equation would be equal to 0.18.[2]

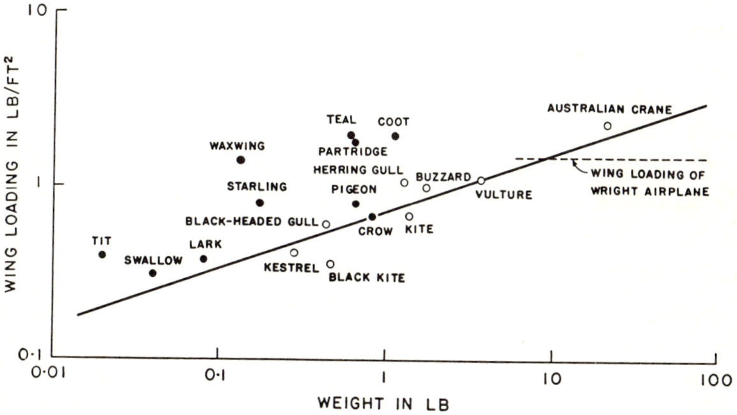

Fig. 10. Wing loading of birds. The wing loading in pounds per square foot is plotted against the weight in pounds, both in logarithmic scale. White circles denote the birds which regularly soar, black circles those which flap their wings. The straight line of slope 1:3 corresponds to the similarity law of Helmholtz.

If we apply the Renard formula to the flight of birds, it is evident that the power required per unit weight of the bird increases with the wing loading. It is interesting to see how the wing loading of birds actually varies with their total weight. Fig. 10 contains information which I have prepared from data given in

[2] Henry gave the formula in the form $P/W = \text{constant} \times \sqrt{W/S}$. In this case the constant is not nondimensional and has the numerical value $\frac{1}{2}$ if P, W, and S are expressed in kilograms, meters, and seconds.

La Machine animale, the famous book written by the French physiologist Etienne Jules Marey (1830–1904) (Ref. 10). The abscissa is the weight in pounds and the ordinate is the wing loading in pounds per square foot, both plotted against logarithmic scales. Distinction is made between birds which regularly soar and those which flap their wings.[3] It is seen that, in general, the wing loading increases as the weight increases. Since we are inclined to believe that the power which a bird can exert by its pectoral muscles is approximately proportional to its weight, it follows that flying becomes more of a problem for a large bird than for a small one. We therefore conclude that there is a certain size beyond which a living being is unable to fly.

The famous German physicist Hermann von Helmholtz (1821–1894) considered the similarity law of flying animals in a paper published in 1873 (Ref. 11). He suggested that the weight of the animal is proportional to the cube and its wing area to the square of its linear dimension. According to this assumption, the wing loading increases proportionally to the cubic root of the weight. This relation is represented by a straight line of slope of 1:3 in Fig. 10, where the logarithmic scale is used. Thus the specific law proposed by Helmholtz seems to be substantiated if we consider only the soaring birds.

In German academic circles an anecdote made the rounds saying that a student failed in an examination held by Helmholtz because he was not able to prove that human flight would never be possible. I doubt that the story is true in this form. Probably the student had been asked about the possibility of flight by man based on human muscular power. Helmholtz came to the conclusion, by considering the influence of increasing weight on flying ability in the animal kingdom, that man has a very poor chance of flight by his own muscle power.

[3] Identification of the birds mentioned in Marey's work and classification of them as soaring and flapping birds were made by Professor Arthur A. Allen, Laboratory of Ornithology, Cornell University, to whom the author extends his most sincere thanks.

Up to now no attempt to operate an airplane by human muscle power has been successful. In 1937 the Italians Bossi and Bonomi succeeded in maintaining level flight of a propeller-driven airplane over a distance of about 2,600 feet, while the propellers were operated by muscle power alone. However, the aircraft was unable to take off under muscle power. Some people believe that, by improving the aerodynamics of wings and propulsion and by reducing the structural weight, an aircraft could be designed to operate by muscle power. A Figure 8 was flown at end of 1977 in California.

In addition to the careful study of bird flight, early researchers in aerodynamics were especially concerned with finding the most favorable wing shapes. Such research was done either in wind tunnels or by means of actual flying in gliders. Fig. 11 shows a series of wing profiles investigated in Phillips' wind tunnel (Ref. 12). We notice that Phillips investigated curved surfaces, which were found more advantageous than flat plates. This observation was fully confirmed by the gliding experiments of Otto Lilienthal (1848–1896) (Ref. 13). Two findings appeared important to the investigators of this period: first, that a curved surface shows positive lift in the case of zero angle of attack, i.e., when leading and trailing edges are located at the same height; second, that the lift-drag ratio of curved surfaces in certain cases is superior to that of flat plates. At that time no theoretical explanation was available of why the curved surface produced lift at zero-angle attitude. We will see later how the modern theory of lift successfully explains this. It is remarkable,

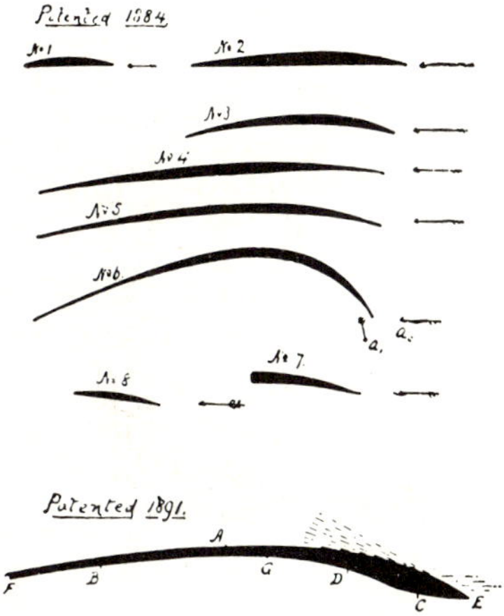

Fig. 11. Wing sections studied by Horatio Phillips. (From *American Engineer and Railroad Journal*, 67 (1893), 135.)

however, to find at a relatively late date (1910) the following comment in the well-known book of Richard Ferris, *How It Flies:* "Later investigations"—he is discussing Henson's airplane design of 1843—"have proven that the upper surface of the aeroplane must be convex to gain the lifting effect. This is one of the paradoxes of flying planes which no one has been able to explain."

Lilienthal strongly emphasized the importance of curved wing surfaces. He made many other interesting aerodynamic observations; he found, for example, that natural wind is more favorable for soaring flight than a perfectly uniform airflow. This favorable effect can be achieved by utilizing the upward components which often exist in the natural wind. Lilienthal found, however, that sometimes the lift in natural wind, even in the absence of upward components, may be superior to that measured in a uniform airstream. Only in recent times was it recognized that this effect is due to a cross-velocity gradient, which generally prevails in the natural wind, at least in lower layers of the atmosphere.

Some theoretical ideas of the Lilienthal brothers, Otto and Gustav (1849–1933), were rather nebulous. They devoted much thought to the possibility of creating negative drag, i.e., propulsion, by a particular shape of the wing profile without providing power. Several years after the death of his brother Otto, who died in an accident in 1896, Gustav Lilienthal actually published a "theory" for this effect, which is evidently in contradiction to the principles of mechanics. In my youthful zeal for scientific truth, I quoted him once as the "small brother of a great man," an expression which I believe hurt him. I regret it now as I look back to this adolescent period of aerodynamic science.

In this country, Octave Chanute (1832–1910), a distinguished civil engineer of Chicago, carried out a great number of gliding experiments. His attention was centered on the problem of stability. It is interesting to know that a month before Otto Lilienthal's fatal accident, he expressed the opinion that Lilienthal's glider was unsafe (Ref. 14).

In addition to man-carrying gliders, flying models with or without propulsion contributed essential aerodynamic data. The model presented by Alphonse Pénaud (1850–1880) seems to have been the first model successfully stabilized by a horizontal tail surface at the rear (Fig. 12). It had a propeller driven by rubber bands. Pénaud thought that a passenger-carrying airplane with a total weight of 2,600 pounds and an engine of 20 to 30 horsepower could be designed in accordance with his patents. His life and work are a tragic chapter in the history of aeronautics. He became paralyzed so that he could only continue his studies in his home; poverty, ill health, and lack of recognition discouraged him to such an extent that he killed himself at the age of thirty.

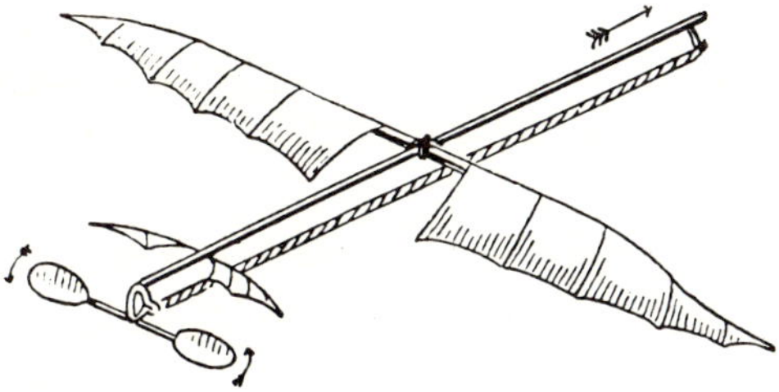

Fig. 12. Alphonse Pénaud's model airplane. (From *American Engineer and Railroad Journal, 66* (1892), 508.)

The Wright brothers, who performed the first mechanical flight of a piloted airplane, and Samuel P. Langley (1834–1906), who came near to such a practical result, followed along the lines we have indicated in this short sketch. Langley strongly stressed the analogy with bird flight and was fully aware that Newton's theory of air resistance could not be correct if human flight with heavier-than-air craft was possible. After flying a power-driven model, he reached the decision to build a man-carrying machine. He was fortunate in possessing as an assistant a mechanical genius who has seldom received the credit due him. This assistant was Charles M. Manly (1876–1927), a Cornell University graduate, who built a gasoline engine of sufficient power and lightness to serve the purpose.

Wilbur (1867–1912) and Orville (1871–1948) Wright were not professional scientists. They were, however, familiar with the practical aerodynamical ideas developed before them by various researchers, and in addition to a remarkable talent for construction, they had the ability to utilize model experiments for their full-scale design. As a matter of fact, they operated a simple and small-scale wind tunnel for this purpose. Furthermore, they carried out nearly one thousand gliding flights.

It is not without interest to consider the characteristic data of the first airplane of the Wright brothers in the light of the theoretical speculations sketched above. The gross weight of their airplane was equal to 750 pounds and the wing had a total area of 500 square feet, so that the wing loading was 1.5 pounds per square foot. This wing loading is a little larger than that of a vulture (Fig. 10), and seventeen times less than that of a fully loaded Douglas DC–3 airplane, for example. The net power available from their 12-horsepower engine with the 66-percent propeller efficiency stated by Orville Wright can be estimated to be 4,300 foot-pounds per second. Hence the power available per unit weight was equal to 5.7 feet per second. According to Renard's formula, the value of the power required per unit weight would be 4.4 feet per second for the above-stated wing loading. It is also interesting to note that Renard, in a paper published in January 1903 (Ref. 15), computed that the engine of a piloted airplane should not be heavier than 17 pounds per horsepower. The engine used by the Wright brothers was 15 pounds per horsepower.

In the year preceding the Wright brothers' first successful flights, the German applied mathematician Sebastian Finsterwalder (1862–1951), published an excellent review of the state of aerodynamic knowledge at that time (Ref. 16). This article contains much interesting material and a great number of references concerning the subject, which I have only been able to touch upon here in a sketchy way.

Hermann von Helmholtz

Gustav Kirchhoff

Lord Rayleigh

Frederick W. Lanchester

(Courtesy of the Royal Aeronautical Society)

CHAPTER II

» *The Theory of Lift*

AS I pointed out in Chapter I, a gap separated the theoretical calculations and the actual observations concerning the magnitude of the lifting capacity of an inclined surface. I also pointed out that, at the time of the first human flight, no theory existed that would explain the sustentation obtained by means of a curved surface at zero angle of inclination of the chord. It seemed that the mathematical theory of fluid motion was unable to explain the fundamental facts revealed by experimental aerodynamics.

There were, however, theoretical results and empirical observations, made independently of the problem of airplane flight, which eventually led to a rational understanding of the phenomena of aerodynamic lift.

Circulation and Magnus Effect

In 1878 Lord Rayleigh, who has been mentioned before, studied the flow around a circular cylinder (Ref. 1). He found that, if the cylinder is exposed to a parallel uniform flow or moves uniformly through a fluid at rest, D'Alembert's theorem applies, and there is no force acting on the cylinder. But the superposition of a circulatory flow upon a parallel uniform flow produces a force perpendicular to the direction of the original flow, or perpendicular to the direction of the motion of the cylinder. This result was used to explain the so-called Magnus effect, which had been well known to artillerists since the beginning of the nineteenth century. The phenomenon had also been recognized by tennis players and golf "duffers." As a matter of fact, Rayleigh's study was undertaken to elucidate the swerving flight of a "cut" tennis ball.

The explanation of this phenomenon is comparatively simple. We start with the theorem of Daniel Bernoulli (1700–1782), which states that in the flow of an incompressible fluid—if for the moment we disregard gravity and frictional effects—the sum of the pressure head and velocity head is constant along a streamline. The pressure head of the stream is the height of a fluid column which, at rest, would produce, by virtue of its weight, the pressure measured in the stream. The velocity head is the height of a fluid column which would produce the same stream velocity through a hole located at the bottom of the column. For example, if an incompressible fluid passes through a horizontal pipe of variable cross section, then because the same fluid mass must go through all the cross sections, the velocity will be smaller in the larger cross section and greater in the smaller cross section. Now it follows from Bernoulli's theorem that where the velocity is higher, the pressure is lower, and vice versa. Bernoulli's theorem can be considered as an expression of the law of the conservation of energy. One can interpret it as a mutual exchange between potential energy and kinetic energy.

Consider now a flow directed from left to right around a cylinder. As shown in Fig. 15, the streamline pattern is completely symmetrical, so that no net force can arise (D'Alembert's paradox). Let us superimpose on this flow a clockwise circulatory motion around the cylinder (Fig. 16). Then, at point *A* we add the velocity of the circulatory motion to the velocity of the flow and get an increased velocity, whereas at *B* the circulatory velocity is directed against the flow and reduces the velocity. According to Bernoulli's theorem, without the circulatory motion the pressure would have the same value at *A* and *B*, but with the circulatory motion the pressure at *B* is higher than at

A, and this pressure difference gives the upward force. If the circulatory motion had a counterclockwise sense, it is evident that the force would be in the opposite direction.

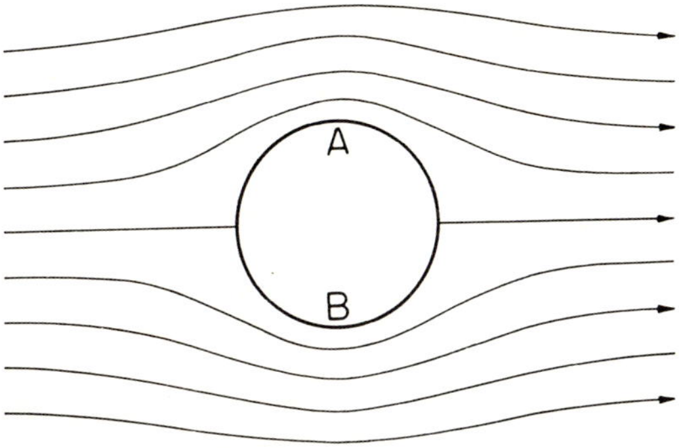

Fig. 15. Ideal flow past a circular cylinder.

Now what happens to the tennis ball can be explained in the following way: The spin given to the ball creates, by friction, a

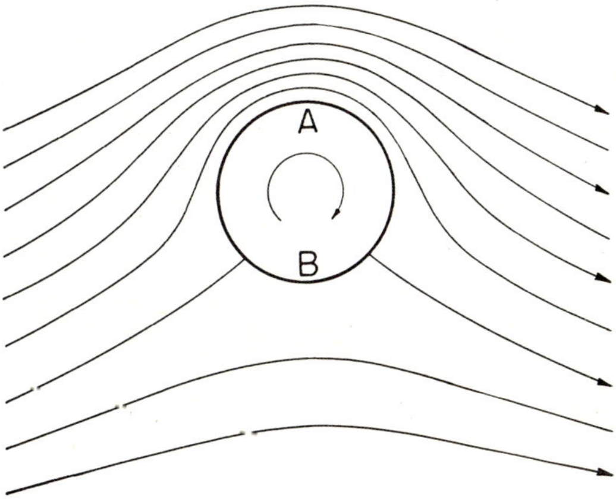

Fig. 16. Ideal flow past a circular cylinder with clockwise circulatory motion superimposed.

circulatory motion of the air in the same direction in which the ball rotates. This circulatory motion, superimposed upon the airflow relative to the ball, produces a force perpendicular to the instantaneous velocity of the ball, i.e., perpendicular to the trajectory of the ball. If the lift is positive, the effect is equivalent to an apparent decrease of gravity; if the lift is negative, it seems to add to the gravity. In the first case the range of the ball is increased, in the second case decreased. The baseball pitcher uses the same effect of spin, directing the ball in a way favorable to his team.

The reader may remember that a German engineer, Anton Flettner (Ref. 2), used the Magnus effect to drive a boat by wind power. If a circular cylinder is erected vertically on a boat and made to rotate around its axis, the wind produces a force in a direction essentially perpendicular to the cylinder axis and the direction of the relative wind. Thus a rotating cylinder can replace a sail. Actually, a sail is nothing but an airfoil. Flettner also showed that by rotating two tandem cylinders in opposite directions he could turn a boat around. I once made a trip out of Bremerhaven, in 1924 I believe, in Flettner's experimental boat. The action of the cylinders as sails was interesting and successful. The ultimate failure of the invention, however, was due to economic reasons. The practical application was intended for cheap freighters or fishing boats. But unfortunately the cylinders had to be driven rather fast in order to obtain significant propulsive forces, and this made necessary the use of ball or roller bearings and the employment of a skilled mechanic for maintenance. The resulting expenses were too high for fishing boats, and the supposed profit in comparison with conventional ship propulsion became illusory. Flettner's experiment was carried out, of course, in a period in which the theory of lift was already well established.

Circulation and Lift: Lanchester, Kutta, and Joukowski

The connection between the lift of airplane wings and the circulatory motion of the air around them was recognized and developed by three persons of very different mentality and training. First I should mention the Englishman Frederick W. Lanchester (1878–1946). He was a practical engineer, more or less an amateur mathematician, and by trade an automobile builder. After working as an engineer in the development of gas engines and producing a new engine starter, he began the construction of the first Lanchester motorcar in 1894. The Lanchester Motor Company, of which he was chief engineer and general manager, was formed in 1899. At the same time he developed the circulation theory of flight, having begun with a paper on that subject in 1894. Two books by him, containing his well-developed ideas, appeared in 1907 and 1908 (Ref. 3). I remember that I visited him in the summer of 1912—on the occasion of the Fifth International Congress of Mathematicians—in England. We met in Cambridge and he showed me about, driving his own car along the narrow English roads at a speed that was rather frightening. This was in the early days of automobiles, and I felt a little uneasy discussing aerodynamics at such speed, but it did not seem to affect Lanchester. He was multisided and full of imagination. For example, during the First World War, he published his ideas on the theory of warfare. A few years ago I found that the first American book on the military science called Operational Analysis starts with a theory of Lanchester's. He was a man who contributed many things to many branches of applied mathematics and who continued to produce technical inventions all his life.

The second person is the German mathematician M. Wilhelm Kutta (1867–1944), who started out as a pure mathematician but became interested in Otto Lilienthal's gliding experiments and therefore in aerodynamic theory. His particular aim was to understand the effect of curvature—why a horizontally placed curved surface produces a positive lift. He published a paper on this subject in 1902 (Ref. 4).

Finally, the third person I should name is Nikolai E. Joukowski, who has been mentioned earlier. He had extensive training in mathematics and physics, obtained originally in Russia and later in Paris. In 1872 he became professor of mechanics at the Polytechnical Institute and in 1886 at the University of Moscow. He had a broad interest in the entire field of theoretical and applied mechanics. In the period 1902–1909, independently of Kutta and Lanchester, he developed the mathematical foundation of the theory of lift, at least for two-dimensional flow, i.e., for wings of infinite span and constant cross section (Ref. 5). As has been mentioned in Chapter I, he was also instrumental in developing the means for aerodynamic research in his country.

Each of these three men recognized the connection between aerodynamic lift and circulatory motion. In order to get a clear concept of the theory of lift, however, we must review some of the fundamental notions of fluid mechanics.

Some Fundamental Notions of Fluid Mechanics: Joukowski's Theorem

If we want to describe the history of a fluid element in a flow, we can show that in the most general case it consists of a translation, a rotation, and a distortion (Fig. 17). In the theory of

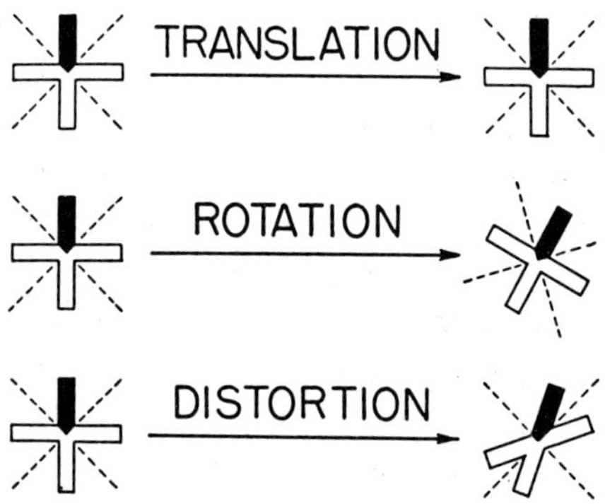

Fig. 17. Translation, rotation, and distortion of a fluid element.

fluid mechanics we call a fluid motion in which the rotation is zero, so that the element is only translated and distorted, a *potential flow* or a *vortex-free flow;* whereas, if the element also rotates, we call the flow a *rotational flow* or a *vortex flow*. The term potential flow originates from the mathematical concept of the velocity potential.

Let us consider a few simple examples of vortex-free and vortex flows. First let us take a parallel flow with uniform velocity. This is evidently the simplest example of a vortex-free flow, because the fluid elements undergo neither rotation nor distortion. All the elements just travel parallel, like automobiles in traffic on a straight road. Second, we consider a two-dimensional parallel *shear flow*, i.e., a flow in which the velocities of all particles are parallel but their distribution through a section perpendicular to the flow direction is nonuniform. This is an example of flow with rotation, or vortex flow. We can explain the concept of rotation in the following way: We place two arrows at a point *A* in a flow which has a linear velocity distribution (Fig. 18),

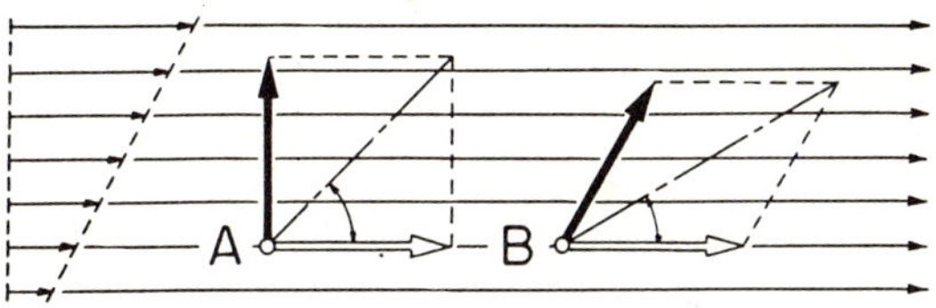

Fig. 18. Parallel shear flow.

one in the stream direction and the other perpendicular to this direction; we observe what happens to the two arrows if they move with the fluid from *A* to *B*. The first arrow is translated parallel to its direction, but the second arrow turns with the stream. In this case we have both distortion and rotation of the element. The magnitude of the rotation of the fluid element is given by the average rotation of the two arrows, i.e., the rotation of their bisectrix. We see that the element rotates, because the angle of inclination of the bisectrix relative to the flow direction, which originally was equal to 45°, decreases as we proceed downstream. This is the simplest example of a vortex flow. It must be noted, however, that the word vortex does not necessarily imply rotation of the whole fluid. Our example is a parallel flow where every element rotates, and, scientifically, it is the rotation of the elements that characterizes a vortex flow. The layman believes that if we talk of vortex flow we must mean that something is whirling around at great speed.

Now let us consider a so-called circulatory flow, in particular a flow in which the fluid elements move around in circular streamlines. If we imagine that the fluid rotates like a rigid body (for example, like a solid wheel), it is clear that we have a vortex flow, because—applying the rule of the two arrows to this case—every element turns around with a certain angular velocity (Fig. 19). There is no distortion. This is the simplest example of a vortex flow with circular streamlines; the angular velocity of the elements is a constant. We call this flow a vortex flow with constant rotation, or *constant vorticity*. Unfortunately, the angular velocity is not the same as vorticity. The two quantities differ by a factor of two because the mathematicians defined the vorticity as twice the angular velocity to give a more esthetic appearance to certain formulas in vector analysis.

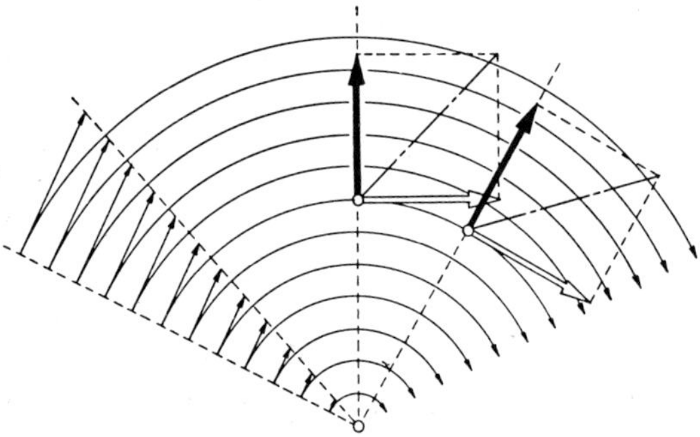

Fig. 19. Circulatory flow with constant vorticity.

Now the question arises whether or not there is a velocity distribution where the streamlines are circles but the flow is vortex-free and the fluid elements do not rotate. The existence of such a flow can be demonstrated, as was the vortex flow, by the use of two arrows. The problem is to find a velocity distribution along a radius such that the bisectrix between the two arrows keeps its original direction. In this case the velocity of the fluid particles necessarily decreases with increasing distance from the center of the circulatory motion. A simple calculation—or an experiment carried out according to the construction indicated in Fig. 20—will readily show that the velocity must be inversely proportional to the distance from the center, *O*. Or we can say that the product $u \times r$ is a constant. In fluid mechanics we prefer to write the formula in the form $u \times 2\pi r$. The expression $2\pi r$ is equal to the circumference of the circular streamline, and the product of the velocity and the circumference is called the *circulation*. So the dimension of circulation is feet per second multiplied by feet.

If the flow is a potential motion, i.e., a vortex-free motion, the circulation is a constant for the whole field of flow. It is evident that such a motion cannot be physically true up to the center, because the velocity at that point would be infinite. So there must be a core or nucleus where the flow is not potential flow. There are two physical possibilities. One possibility is that in the nucleus we have fluid that rotates. We usually assume that the nucleus rotates approximately like a solid body—i.e., that the vorticity has a constant value within the nucleus (Fig. 20). Such a combination we call a *vortex* or an *eddy*. It consists of a fluid

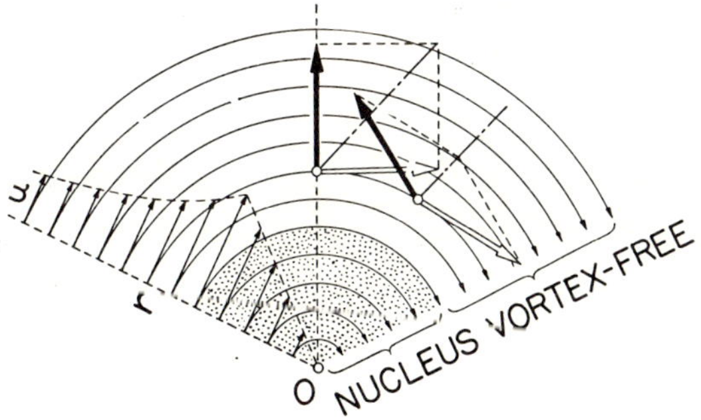

Fig. 20. Circulatory flow with nucleus inside and vortex-free flow outside. The center is at *O*; *u* denotes the fluid velocity (tangential) and *r* the radius.

nucleus rotating like a solid body and a circulatory flow with outward decreasing velocity. However, instead of a fluid nucleus we can also have a solid body, as core. Then outside the solid body we may have a circulatory flow without vorticity. This is the case that we are considering, for example, when we talk of the Magnus effect. We assume first that a circulatory flow exists around a ball or a cylinder. Then we give the body a translatory motion, and the combined flow produces lift. Joukowski has shown that when a cylindrical body of arbitrary cross section moves with the velocity U in a fluid whose density is ρ and there is a circulation of the magnitude Γ around it, a force is produced equal to the product $\rho U \Gamma$ per unit length of the cylinder. The direction of the force is normal both to the velocity U and the axis of the cylinder.

So we have an explanation of the lift phenomenon if we can show that there is really circulation around the body. For the reader who likes to think in mathematical or geometrical terms, I will note that one can generalize the definition of circulation by taking the mean value of the component of the velocity along an arbitrary closed curve encircling a body and multiplying it by the length of arc of that curve. If the flow is vortex-free, this product has the same value independent of the choice of the curve. Thus we have a general definition of circulation, generalized from a circulatory flow with circular streamlines. If we take a closed curve that does not go around the body but encloses fluid only, then the circulation around the curve will be zero.

Two-Dimensional Wing Theory (*Wing of Infinite Span*)

For the lift problem, as far as an infinite wing of constant section is concerned, we assume that the flow around the wing is vortex-free. Then the computation of the lift is reduced to the determination of the magnitude of the circulation as a function of the velocity and as a function of the shape of the wing section.

This problem was solved in principle by Kutta and by Joukowski. The best way to arrive at an understanding of their solution is to consider the flow pattern around a wing section put into motion in a fluid originally at rest.

First I must mention a fundamental theorem on vortex motion announced by Helmholtz (Ref. 6). This great German physicist showed that if there is no initial vorticity in a fluid, e.g., if the fluid is originally at rest, vorticity can only be created by friction or by the presence of sharp edges on a body. In the latter case, a discontinuity may be formed between two fluid streams meeting at the edge. Fig. 13 (p. 26) shows, for example, a discontinuity between a fluid in motion and a fluid at rest. Such a discontinuity can be considered as a continuous sequence of vortices, or a *vortex sheet*.

We now want to observe what happens when a wing section with a sharp trailing edge is put in motion. (We call the front part of the wing, exposed to the stream, the leading edge and the part in the rear, where the fluid stream leaves the wing surface, the trailing edge.) The leading edge is usually rounded, at least for wings used at subsonic speeds, whereas the trailing edge is made as sharp as possible. Figs. 21 and 22 show flow photographs in which the streamlines are made visible by the introduction of

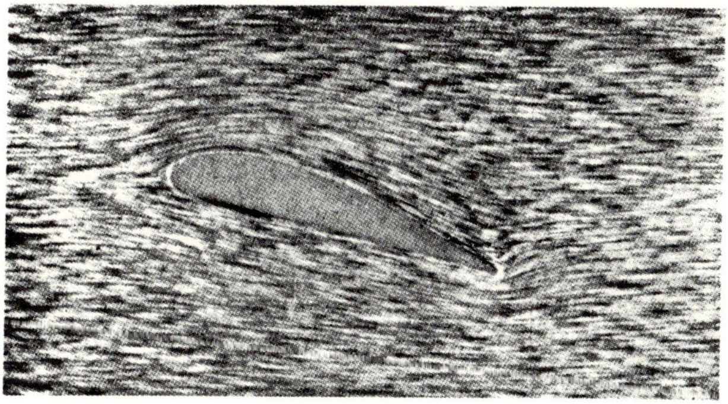

Fig. 21. Picture of the streamline flow around an airfoil started from rest. The camera is moving with the airfoil. (From L. Prandtl and O. G. Tietjens, *Applied Hydro- and Aeromechanics* [copyright 1934, United Engineering Trustees, Inc., McGraw-Hill Book Co., Inc.], by permission.)

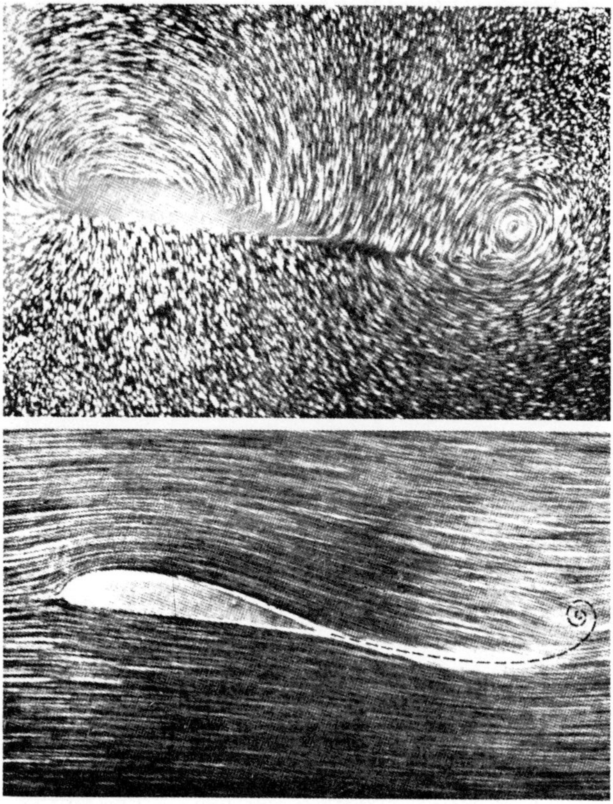

Fig. 22. Pictures of the streamline flow at a little later stage than Fig. 21. *Upper:* Camera at rest relative to undisturbed fluid. *Lower:* Camera moving with the airfoil. (From L. Prandtl and O. G. Tietjens, *Applied Hydro- and Aeromechanics* [copyright 1934, United Engineering Trustees, Inc., McGraw-Hill Book Co., Inc.], by permission.)

fine aluminum powder, which supposedly follows the streamlines of the fluid. We see that at the first moment, as shown in Fig. 21, the fluid has the tendency to "go around" the sharp edge. However, we may say that the fluid does not like this process, because a very high (theoretically infinite) velocity is required at the edge. Instead, a vortex is created at the sharp edge, and it is followed by a discontinuity, or vortex sheet. Now we must remember that, according to a fundamental principle of mechanics, a rotation, or more exactly a moment of momentum, cannot be created in a system without reaction. For example, if we try to put into rotation a body, such as a wheel, we experience a reaction tending to rotate us in the opposite direction. Or in the case of a helicopter with one rotor turning in one direction, we need a device to prevent the body of the craft being put into rotation in the opposite sense. Similarly, if the process of putting a wing section in motion creates a vortex, i.e., a rotation of a part of the fluid, a rotation in the opposite sense is created in the rest of the fluid. This rotary motion of the fluid appears as the circulation around the wing section. In a way analogous to what we have seen in the case of the tennis ball, the circulation creates higher velocity (lower pressure) at the upper, and lower velocity (higher pressure) at the lower, surface of the wing. In this manner a positive lift is produced.

It is clear that this point of view changes the whole physical picture concerning lift. In earlier times the instinctive impression was that the air hits the inclined wing surface and that the airplane is therefore carried by the air below. We now see that the airplane wing is at least partially hung up or sucked up by the air passing along the upper surface. As a matter of fact, the contribution to the total lift from the negative pressure or suction developed at the upper surface is larger than the contribution from the positive pressure at the lower surface.

Let us return to the process of the development of circulation. We saw that a vortex is created near the trailing edge; it is left behind as the wing proceeds. We call this vortex the starting vortex. It can be clearly seen in the photographs of Fig. 22. Simultaneously, as we mentioned above, a circulation is generated around the wing section, and as long as vorticity leaves the wing in the vortex sheet, the circulation increases. However, it is reasonable to assume that, when the starting vortex is swept far away, the circulation has reached its maximum value, as there is no longer a velocity difference between the flows leaving the upper and lower surfaces. This assumption was put forward independently by Kutta and Joukowski. It is called the Kutta-Joukowski condition, or the condition of smooth flow at the trailing edge. It is the salient point in the theory of lift because it determines the magnitude of the circulation. By means of this hypothesis the whole problem of lift becomes purely mathematical: one has only to determine the amount of circulation so that the velocity of the flow leaving the upper surface at the trailing edge is equal to that of the flow leaving the lower surface. The rule stated in this way applies to wings with zero vertex angle at the trailing edge. If the tangents to the upper and lower surfaces form a finite angle, the trailing edge is a stagnation point, i.e., the velocity computed from both sides is zero.

The Kutta-Joukowski condition seems to be a reasonable hypothesis, both because it is indicated by visual observation and also because the lift calculated by means of this condition is in fair accordance with measurements. A comparison between theory and experiment is shown in Fig. 23, where the lift co-

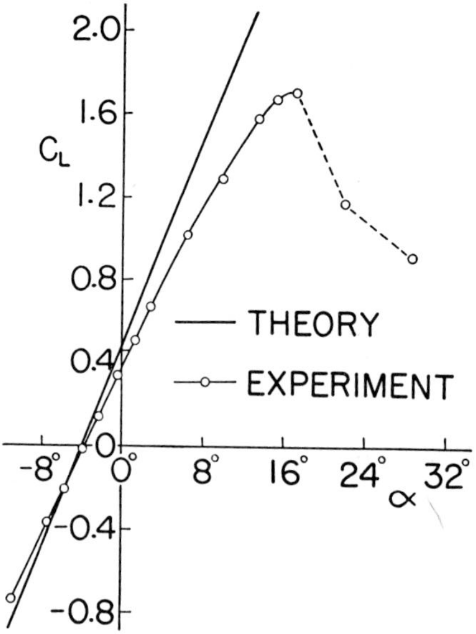

Fig. 23. Lift coefficient C_L of an N.A.C.A. 4412 airfoil versus angle of attack, α. The circulation theory of list is compared with the experimental result.

efficient is plotted against the angle of attack, α, for a typical wing section. The lift coefficient, C_L, is a nondimensional quantity obtained by dividing the lift force per unit width by the chord length, L, and by the dynamic pressure, $\frac{1}{2}\rho U^2$, where ρ is the density of fluid and U is the velocity of flight or the velocity of the undisturbed flow relative to the wing. The curve obtained by calculation agrees fairly well with measured values provided that the angle of attack is not large. The calculated pressure distribution for the same wing section is also compared with the measured result in Fig. 24, where the difference between the

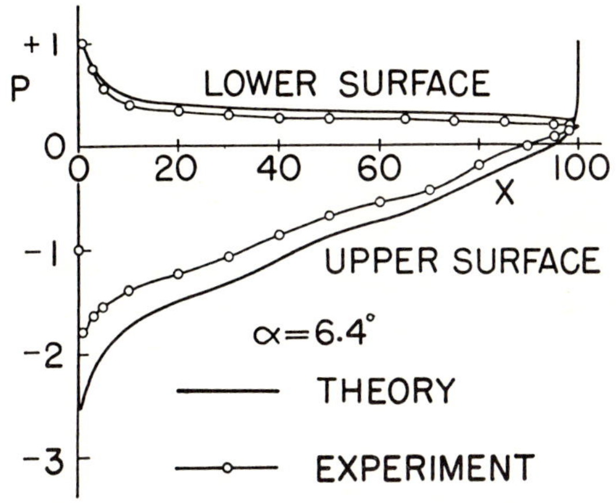

Fig. 24. Pressure distribution along the chord of an N.A.C.A. 4412 airfoil at an angle of attack $\alpha = 6.4°$. P is the pressure on the surface relative to that of the stream, divided by the dynamic pressure of the stream, and X is the distance along the chord in percentage of the chord. The circulation theory of lift is compared with the experimental result.

pressure acting on the surface (both upper and lower) and the pressure prevailing in the undisturbed flow, divided by the dynamic pressure, is plotted. Again the agreement between theory and experiment is good.

Here I want to point out that the result of this theory, which we call the circulation theory of lift, differs considerably from Newton's theory. In Newton's theory it is assumed that the air mass which is deflected is the amount of air which directly hits the surface of the body. If the chord of the flat plate is L and the angle of attack is α, then the mass of air which is deflected per unit width of the plate is proportional to $L \sin \alpha$ (Fig. 5, p. 10). According to the circulation theory, however, it is proportional to 3.14 L. If α is 5°, for example, $\sin \alpha$ being less than 0.1, Newton's result is in error by more than a factor of 30. A comparison between Newton's theory and the circulation theory is also seen in Fig. 14 (p. 26), which shows the non-dimensional normal force (i.e., the force component normal to the plate; whereas the lift is the component normal to the direction of the relative stream) plotted against the angle of attack.

Limitation of the Wing Theory: Stall

Fig. 23 shows that the usefulness of the theory is restricted to a limited range of angle of attack, comprising relatively small angles, positive and negative. Beyond this range the measured lift falls far below the values predicted by the theory. The physical explanation of this discrepancy—supported by visual observation—is the following: As already mentioned, the lift acting on a wing is due to a difference in pressure between the upper and lower surfaces. This difference in pressure can only be maintained if the flow follows the surface. At small angles of attack the flow has little difficulty in following the surface. As the angle is increased, however, the air finds it increasingly difficult to maintain contact, especially on the upper surface, where it has to work its way against increasing pressure, and it separates from the surface before reaching the trailing edge. The separation results, first, in a considerably lower value of circulation than that which the Kutta-Joukowski condition prescribes and, second, in an actual decrease of circulation with an increasing angle of attack. Thus there exists a certain critical angle of attack for every wing section, beyond which the lift no longer increases with the angle but starts to decrease. Then the wing is said to be *stalled.* This phenomenon has great importance, because it determines the maximum load that the wing can sustain at a given speed and, in particular, the safe landing speed of an airplane.

Stall also sometimes appears to occur to birds. One can make a bird stall if one has some knowledge of aerodynamics. I often tried to do this with seagulls on the shore of Lake Constance. I had bread in my hand and as the birds tried to get it, I slowly withdrew my hand. Then the birds tried to decrease their speed to get it, which required an increased lift coefficient. Several times, apparently, the birds exceeded the critical angle of their wings and stalled. The difference between the bird and an airplane is that the bird can easily produce additional lift by vigorously flapping its wings.

The phenomenon of flow separation depends largely on viscous effects, which are neglected in the circulation theory of lift. We do not yet have a reliable theory to predict the angle at which a stall will occur nor the flow pattern around the wing when it is stalled. We know, however, certain means which are effective not to prevent completely, but to postpone, the stall. Such means are called *high-lift devices.*

One such device is a slot near the leading edge, an invention of Gustav Lachmann and Sir Frederick Handley-Page. The slot prevents flow separation from the neighborhood of the leading edge, which is the most dangerous type of separation. Instead of a fixed slot in the wing, one can also arrange a movable winglet ahead of the leading edge. The winglet is moved ahead automatically by the negative pressure at high angles of attack and creates a slot which is kept closed in normal flight. Lachmann, a German pilot in the First World War who later earned his doctor's degree at the University of Aachen with a thesis on the theory of the slot, told me that he conceived the idea of the slot while in a military hospital after a serious accident due to stalling. Handley-Page in England independently arrived at the same invention. Later they worked together. Another device for delaying stall is the split flap or the slotted flap fitted near the trailing edge. Almost every airplane has such a trailing-edge flap, and you can watch it in operation during landings.

Three-Dimensional Wing Theory (*Wing of Finite Span*)

Of the three men I have named as pioneers in circulation theory, only Lanchester went further than the problem of a wing of infinite span with constant section. He was the first man to attack the problem of a wing of finite span. He had the idea that, if a wing, by its motion, creates a circulation around itself—what he called a "peripteral motion"—then it must really behave as a vortex, i.e., it must induce a flow field just as a vortex segment of the length of the span would do. So he replaced the wing by a *bound vortex*, "bound" meaning that it cannot swim freely in the air like a smoke ring but moves with the wing. Its core is the wing itself. According to the Helmholtz theorem (Ref. 6), however, a vortex cannot begin or terminate in the air: it must end at a wall or form a closed loop. So Lanchester concluded that, if the bound vortex ends at the tip of the wing, there must be some continuation, and this continuation must be a *free vortex*—"free" because it is no longer confined to the wing. Therefore, the wing can be replaced by a vortex system consisting of a bound vortex which travels with the wing and free vortices springing from the wing tips and extending downstream. Lanchester recognized this basic fact, as is shown by his drawing of the vortex system reproduced in Fig. 25.

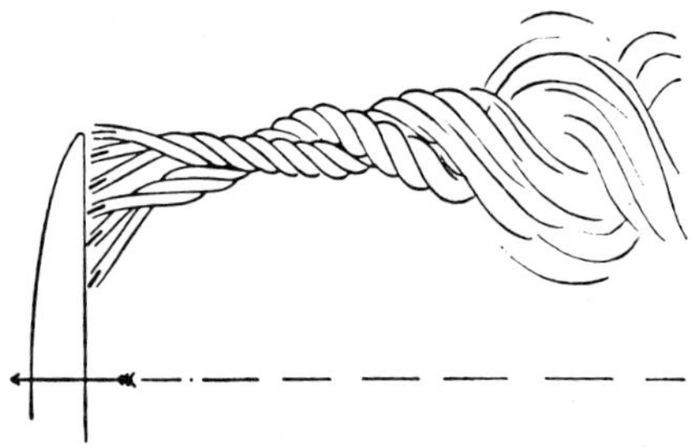

Fig. 25. Lanchester's drawing of the vortex system around a wing. (From F. W. Lanchester, *Aerodynamics* [London, 1907], by permission of Constable and Co., Ltd.)

Sometimes one can see the tip vortices when they are made visible by condensation trails. The air is sucked into the low-pressure vortex core and is cooled by thermal expansion to such a degree that the water vapor contained in the air condenses. Fig. 26 is a picture of an airplane flying over a forest and emitting insecticide dust from its trailing edge. In this case one clearly sees the edges of the dust sheet rolled up by the rotation induced by the tip vortices.

Fig. 26. An airplane flying over a forest and emitting dust. (Courtesy of U.S. Forest Service.)

The system of free vortices gives rise to a velocity field, called the field of induced velocities, in which each constituent vortex with horizontal axis sets up a circulatory motion of air. We are especially interested in the vertical component of the velocity in this field, which we call the *downwash*. According to general mechanical principles, every force acting on a body moving through the air must have its counterpart in the momentum imparted to the air. Thus the lift gives rise to a downward motion of the air behind the airplane; this is the downwash. As the airplane proceeds, new air masses are pushed downward and the momentum created in unit time is equal to the lift force.

This concept also gives the correct answer to the age-old question concerning the energy required for sustentation. I referred in Chapter I to early speculative calculations of the amount of work necessary for sustentation. Lanchester, however, was the first man to point out that the kinetic energy of the downwash field represents the work necessary to obtain sustentation. One important consequence is that no such work would be necessary if the wing were infinitely long. If we compare two wings with the same lift and the same area but with different spans, we find that the work is less for the longer wing than for the shorter. The ratio between the span and the mean chord is called the *aspect ratio*. Lanchester was the first to recognize the importance of the aspect ratio of the wing in connection with the work required for sustentation.

Lanchester and Prandtl

The man who gave modern wing theory its practical mathematical form was one of the most prominent representatives of the science of mechanics, and especially fluid mechanics, of all time, Ludwig Prandtl (1875–1953). He was my teacher at Göttingen University; I was his assistant. His greatest contributions to fluid mechanics were in the field of wing theory and the theory of the boundary layer, of which I will speak in the next chapter.

Prandtl, an engineer by training, was endowed with rare vision for the understanding of physical phenomena and unusual ability in putting them into relatively simple mathematical form. His control of mathematical methods and tricks was limited;

many of his collaborators and followers surpassed him in solving difficult mathematical problems. But his ability to establish systems of simplified equations which expressed the essential physical relations and dropped the nonessentials was unique, I believe, even when compared with his great predecessors in the field of mechanics—men like Leonhard Euler (1707–1783) and D'Alembert. He obtained much of his training from August Föppl (1854–1924) in Munich. Föppl himself did pioneering work in bringing together applied and theoretical mechanics. Later Prandtl became Föppl's son-in-law, following the good German academic tradition. There is a saying—I do not know its author—that it is remarkable how often scientific talent in Germany has descended from father-in-law to son-in-law, instead of from father to son!

There has been some discussion in the literature of how much credit was due to Lanchester and how much to Prandtl for the development of modern wing theory. Lanchester, at the end of his life, was quite embittered because he felt his contributions were not adequately recognized. Everyone talked only of "Prandtl vortices" and the "Prandtl wing theory." I remember that Lanchester came to Göttingen long before Prandtl published his wing theory—at the time I was a graduate student there—and explained many ideas which he published later. Both Prandtl and Carl Runge (1856–1927) were present and learned very much from these discussions. Runge was professor of applied mathematics in Göttingen and acted as interpreter, because neither Lanchester nor Prandtl could speak the other's language. Some feel that Prandtl in his publications did not give full recognition to Lanchester, as far as priority of ideas is concerned.

Many great men who have the imagination to work out systems of ideas for themselves share the weakness of forgetting where an early inspiration came from. For example, it seems that Sir Joseph John Thomson (1856–1940), the great English physicist, was somewhat that way. His pupil and friend, Sir Francis W. Aston, once told me that it was quite amusing to tell something new to Thomson. If you told him an idea on Wednesday, he shook his head; on Thursday, again he would not believe it; but on the next Monday, he would come to you saying, "Now, look here, the thing is so . . ." Thereupon he would propound the same idea you had told him before, ending with, "Now, do you understand the problem?"

It is hard for an active and creative brain to remember from what reading or from what conversation the first inspiration arose. So I am sure Prandtl never felt that he did not give full recognition to Lanchester's work. It was probably not quite clear to him how many elements of the theory that he worked out with such great success were already contained in Lanchester's work. Prandtl touched upon this subject on the occasion of his delivery of the 1927 Wilbur Wright Memorial Lecture to the Royal Aeronautical Society (Ref. 7):

In England you refer to it as the Lanchester-Prandtl theory, and quite rightly so, because Lanchester obtained independently an important part of the results. He commenced working on the subject before I did, and this no doubt led people to believe that Lanchester's investigations, as set out in 1907 in his "Aerodynamics," led me to the ideas upon which, the aerofoil theory was based. But this was not the case. The necessary ideas upon which to build up that theory, so far as these ideas are comprised in Lanchester's book, had already occurred to me before I saw the book. In support of this statement, I should like to point out that as a matter of fact we in Germany were better able to understand Lanchester's book when it appeared than you in England. English scientific men, indeed, have been reproached for the fact that they paid no attention to the theories expounded by their own countryman, whereas the Germans studied them closely and derived considerable benefit therefrom. The truth of the matter, however, is that Lanchester's treatment is difficult to follow, since it makes a very great demand on the reader's intuitive perceptions, and only because we had been working on similar lines were we able to grasp Lanchester's meaning at once. At the same time, however, I wish to be distinctly understood that in many particular respects Lanchester worked on different lines than we did, lines which were new to us, and that we were able to draw many useful ideas from his book.[1]

Prandtl's Lifting-Line Theory: Wings with High Aspect Ratio

Prandtl (Ref. 8) systematized the ideas and simplified the picture in the following way: (*a*) the wing is replaced by a lifting line perpendicular to the flight direction; (*b*) the lifting line is assumed to consist of a bound vortex with circulation variable in order to account for the fact that the lift may change along the span; (*c*) in accordance with the change in the circulation along the span, free vortices are born and extend downstream; however, (*d*) the flow produced by the vortex system is considered as a small perturbation of the fundamental stream relative to the wing, and therefore (*e*) it is assumed that the free vortices approximately follow the original direction of the streamlines parallel and opposite to the flight direction, instead of winding up immediately into a tip vortex as Lanchester assumed (Fig. 25); (*f*) the flow in the immediate neighborhood of a wing section is determined by the two-dimensional solution given by Kutta and Joukowski.

With these assumptions the problem of lift becomes accessible to mathematical treatment, whereas the original concept of Lanchester is difficult to express in mathematical form.

Taking into account the change of the relative wind by the induced flow, we obtain from (*f*) the result that the lift of every individual wing element, and also the total lift on the wing, are linear functions of the angle of attack, as in the two-dimensional theory; but the slope of the line of lift versus the angle of attack depends on the aspect ratio and decreases with decreasing aspect ratio. This was already recognized by Lanchester in a qualitative way.

By means of Prandtl's theory, we can solve two problems. First, if the distribution of lift along the wing span is known,

[1] Reproduced by permission of the Royal Aeronautical Society.

we can determine the flow pattern of induced velocities by a straightforward calculation, and also the energy necessary to obtain the lift distribution; second—and this is more interesting for the engineer—we can determine the lift distribution along the span when the geometry of the wing is given, i.e., when the distributions of chord, wing section, and angle of attack along the span are given. The second problem is mathematically somewhat more involved than the first one. It requires the solution of an integral equation—not necessarily a straightforward calculation. Many methods have since been worked out to solve the specific integral equation of the lift problem. There are analytical methods using developments in infinite series, graphical methods, and methods of successive approximation. One of the interesting methods is that developed by Sears (Ref. 9), which starts with an idea which he and I worked up together and uses the method of eigen-functions in the Schmidt-Fredholm manner.

The solution of Prandtl's integral equation gives the designer important information about the influence of such geometrical features of a wing as aspect ratio, chord and twist distributions, and aileron and flap displacements. Thus the wing theory became the very basis of the scientific design of all our airplanes, at least as far as the domain of moderate speeds is concerned.

To be sure, Prandtl's theory has limitations, as does every theory. Its first limitation is caused by the phenomenon of *stall*. This is the same limitation I have already mentioned in the discussion of the two-dimensional theory of Kutta and Joukowski; namely, the magnitude of the circulation cannot be predicted theoretically when the angle of attack exceeds a certain limit, because the flow separates from the surface.

The second specific limitation of the lifting-line theory concerns the *sweepback* of the wing, a feature which has been adopted for high-speed airplanes for reasons that will be better explained in Chapter IV. If we replace a sweptback wing by a sweptback lifting line, we find difficulty in calculating the downwash because it becomes mathematically infinite on the lifting line.

The third limitation is that the lifting-line theory does not give a good approximation for wings of *low aspect ratio*. If the span is not large in comparison to the mean chord of the wing, it cannot be assumed that the flow pattern in a plane perpendicular to the span is approximated by two-dimensional flow.

Extension of the Lifting-Line Theory: Jones's Theory for Low-Aspect-Ratio Wings

In the latter two cases, i.e., the wing with sweepback and the wing with low aspect ratio, we must proceed to a more exact and more complicated theory in which the wing is represented by a lifting surface instead of a line. The situation is still similar to that which occurs in the case of the lifting line. If we know the lift distribution over the surface, we are able to calculate in a straightforward way the flow field and the energy required to obtain the given sustentation. If, however, the geometric shape of the surface is given, the solution of the problem of determining the lift distribution involves great mathematical difficulties, because we have to solve an integral equation containing a double integral, and in this task even the best mathematicians have not helped us much. We aerodynamicists have had to go ahead with our own devices. Such a theory was initiated first in Prandtl's school by Blenk (Ref. 10). Although a great amount of work has been done on this problem, I cannot go into further discussion here. Perhaps the most systematic presentation of the present status of the question has been given by Flax and Lawrence of the Cornell Aeronautical Laboratory (Ref. 11).

However, I would like to mention an interesting approach toward an approximate solution of the problem for a wing with a very low aspect ratio. The analysis in this case is remarkably simple and is due to an ingenious scientist of the younger generation, Robert T. Jones, who works for the National Advisory Committee for Aeronautics (N.A.C.A.) (Ref. 12).

Jones is an example of a scientist who has made great contributions to aerodynamics without having the advantage of systematic education leading to a degree. As a matter of fact, he was in college only two semesters; thereafter he had various jobs, including trying to build airplanes for a small company. The depression finished this venture, and he found himself operating an elevator in a government building in Washington. I thought Max Munk, himself a leading aerodynamicist, discovered the potential talents of the elevator boy, but Arthur Kantrowitz recently destroyed my story; according to him Jones got a job in the N.A.C.A. on the recommendation of his home-town congressman—a story which is much less dramatic. In any case, Jones's superiors at the N.A.C.A. gave him a chance to continue his studies by reading scientific literature and attending lectures.

Jones considered wings of very small aspect ratio (Fig. 27).

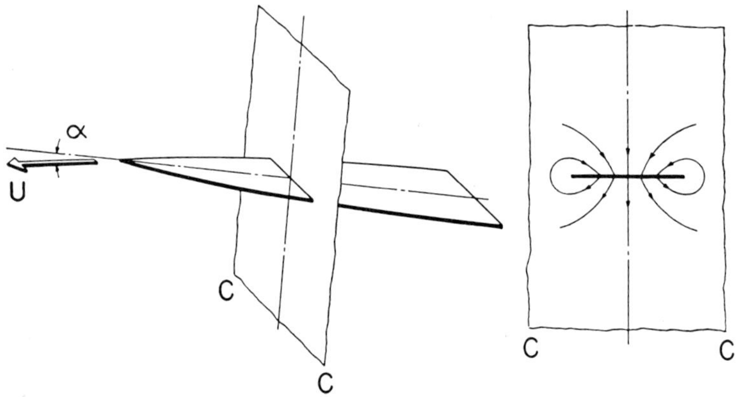

Fig. 27. The wing of low aspect ratio. The flow around every cross section perpendicular to the flight direction can be approximated by the two-dimensional flow around the same cross section. U is the flight speed and α is the angle of attack. C—C represents a typical cross section.

As we shall see in a later chapter, such wings, for example the well-known delta wing, have recently become very important because of their application in high-speed flight. I have mentioned before that Prandtl assumed the flow around every cross section of the wing perpendicular to the span to be approximated by a two-dimensional flow. For wings of very small aspect ratio, Jones made the opposite of Prandtl's assumption. He postulated that the flow around every cross section perpendicular to the flight direction can be approximated by the two-dimensional flow around the same cross section, superposed on the original uniform stream. This idea makes it possible to determine the lift distribution along the chord just as the Prandtl theory gives the lift distribution along the span.

One of the remarkable results of the theory of Jones is the fact that the lift at any point of the chord is only influenced by the flow ahead of the point considered and is independent of the flow conditions downstream, whereas in Prandtl's case of large-aspect-ratio wings, the local lift depends largely on the influence of the free vortices downstream. Jones's theory furnishes an important counterpart to Prandtl's theory and, in my opinion, rounds out the wing theory in a very satisfactory way. I should mention that a similar idea was used before by Munk (Ref. 13) to calculate the forces acting on airship hulls, a problem which is out of date today. Munk did not think of the possibility of applying the same idea to wing theory, whereas Jones recognized the value of such a theory for the solution of a quite modern problem, viz., that of the delta wing.

In this presentation of the theory of lift, I have deliberately left out an important aspect, namely, the influence of viscosity on lift phenomena. As a matter of fact, the influence of viscosity enters in almost every aerodynamic problem when we go into a more complete analysis. However, we shall discuss the viscous phenomena in the next chapter, in connection with the problem of drag, where viscosity has a primary influence.

References

1. Rayleigh, Lord, "On the Irregular Flight of a Tennis-Ball," *Messenger of Mathematics*, 7 (1878), 14–16; also *Scientific Papers* (Cambridge, 1899), I, 344–346.

2. Flettner, A., "Die Anwendung der Erkenntnisse der Aerodynamik zum Windantrieb von Schiffen," *Werft, Reederei, Hafen*, *5* (1924), 657–667.

3. Lanchester, F. W., *Aerodynamics* (London, 1907); *Aerodonetics* (London, 1908).

4. Kutta, M. W., "Auftriebskräfte in strömenden Flüssigkeiten," *Illustrierte Aeronautische Mitteilungen*, *6* (1902), 133–135; "Über eine mit den Grundlagen des Flugproblems in Beziehung stehende zweidimensionale Strömung," *Sitzungsberichte der Bayerischen Akademie der Wissenschaften, mathematisch-physikalische Klasse* (1910), 1–58; "Über ebene Zirkulationsströmungen nebst flugtechnischen Anwendungen," *ibid.* (1911), 65–125.

5. Joukowski, N., "On the Adjunct Vortices" (in Russian), *Obshchestvo liubitelei estestvoznaniia, antropologii i etnografii, Moskva, Izviestiia*, *112*, *Transactions of the Physical Section*, *13* (1907), 12–25; "De la chute dans l'air de corps légers de forme allongée, animés d'un mouvement rotatoire," *Bulletin de l'Institut Aérodynamique de Koutchino*, *1* (1912), 51–65; "Über die Kontouren der Tragflächen der Drachenflieger," *Zeitschrift für Flugtechnik und Motorluftschiffahrt*, *1* (1910), 281–284; *3* (1912), 81–86; *Aérodynamique* (Paris, 1916 and 1931).

6. Helmholtz, H. von, "Über Integrale der hydrodynamischen Gleichungen, welche den Wirbelbewegungen entsprechen," *Journal für die reine und angewandte Mathematik*, *55* (1858), 25–55.

7. Prandtl, L., "The Generation of Vortices in Fluids of Small Viscosity," *Journal of the Royal Aeronautical Society*, *31* (1927), 720–741.

8. Prandtl, L., "Tragflügeltheorie," *Göttinger Nachrichten, mathematisch-physikalische Klasse* (1918), 451–477; (1919), 107–137 (reprinted by L. Prandtl and A. Betz in *Vier Abhandlungen zur Hydrodynamik und Aerodynamik* [Göttingen, 1927]); "Application of Modern Hydrodynamics to Aeronautics," *N.A.C.A. Report* No. 116 (1921).

9. Sears, W. R., "A New Treatment of the Lifting-Line Wing Theory, with Application to Rigid and Elastic Wings," *Quarterly of Applied Mathematics*, *6* (1948), 239–255.

10. Blenk, H., "Der Eindecker als tragende Wirbelfläche," *Zeitschrift für angewandte Mathematik und Mechanik*, *5* (1925), 36–47.

11. Flax, A. H., and Lawrence, H. R., "The Aerodynamics of Low-Aspect-Ratio Wings and Wing-Body Combinations," *Proceedings of the Third Anglo-American Aeronautical Conference* (1951), 363–398.

12. Jones, R. T., "Properties of Low-Aspect-Ratio Pointed Wings at Speeds below and above the Speed of Sound," *N.A.C.A. Report* No. 835 (1946).

13. Munk, M. M., "The Aerodynamic Forces on Airship Hulls," *N.A.C.A. Report* No. 184 (1923).

Ludwig Prandtl

CHAPTER III

» *Theories of Drag and Skin Friction*

IN THE last chapter we considered the lifting power of a wing. All the forces involved were pressure forces on the wing, e.g., positive pressure on its lower, and negative pressure on its upper, surface. We neglected the forces acting tangentially to the surface, which are called *frictional forces.* When we consider drag, we can no longer neglect frictional forces.

Let us analyze all the forces that act on a body moving through a fluid originally at rest. We have *pressure drag* and *frictional drag.* Pressure drag is the component, parallel to the direction of motion of the body, of the force resulting from all the pressures. Frictional drag is the resultant of all the tangential forces taken in the same direction. Pressure drag has its origin in two phenomena. One is related to the lift, that is, to the work which must be expended to obtain lift. The force which necessitates the expenditure of this work is called *induced drag.* The other part of the pressure drag is independent of lift, and I should like to call it *wake drag.*

The induced drag is zero if the span is infinite. In this case, as was shown in the last chapter, no work is required for sustentation; hence there is no induced drag. The wake drag is zero if we neglect friction and assume that the flow closes around the wing, as it is described by the mathematical solution for nonviscous fluids. This is according to the theorem, mentioned earlier, which we called the paradox of D'Alembert. In real fluids, however, because of frictional effects, the streamlines do not follow the surface of the body back to the rear end but separate from the surface somewhere, thus leaving downstream an eddying region called the *wake*. Consequently, the pressure over the rear part of the body cannot reach such high values as are calculated for the nonviscous flow. Because the pressures at the front and at the rear are no longer balanced, a pressure drag occurs. This is the wake drag.

The wake drag and the frictional drag together are called profile drag, because they are determined by the local cross section (profile) of the wing. There are, therefore, two standpoints for classifying drag: one, whether drag comes from pressures or from frictional forces; the other, whether it depends on lift or on the profile of the wing.

Induced Drag

Let us consider these different kinds of drag somewhat more closely. The aeronautical engineer generally uses nondimensional coefficients instead of the forces themselves. For example, the lift coefficient, C_L, already used in Chapter II, and the drag coefficient, C_D, are defined by dividing the lift and drag, respectively, by the wing area and by the dynamic pressure corresponding to the velocity of flight. The dynamic pressure is the amount of pressure increase which appears when the fluid flow of density, ρ, and velocity, U, is brought to rest; it is equal to $\frac{1}{2}\rho U^2$. Fig. 28 is a diagram very familiar to aeronautical engineers, the so-called *polar diagram*, in which the lift coefficient is plotted against the drag coefficient. The angle of attack is used as a parameter. The data are the results of wind-tunnel measurements on wings of aspect ratio ranging from 1 to 7 (Ref. 1). The aspect ratio, as explained in Chapter II, is obtained by dividing the span by the mean chord.

Now according to the lifting-line theory of Prandtl, the coefficient of induced drag is proportional to the square of the lift

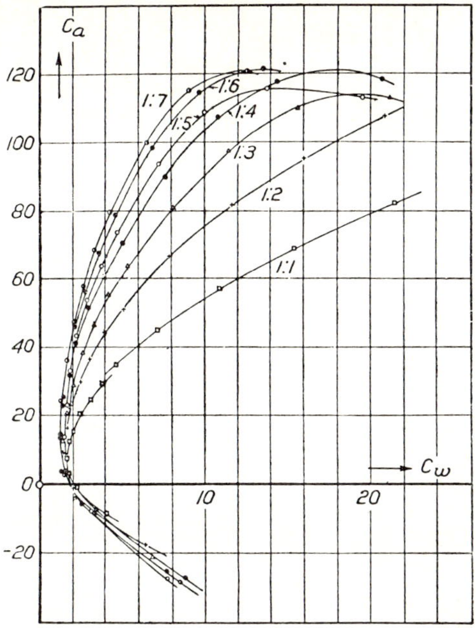

Fig. 28. Experimental values of lift coefficient C_a against drag coefficient C_w for a series of wings of different aspect ratios. The numbers on the curves refer to the aspect ratios. (From L. Prandtl, "Application of Modern Hydrodynamics to Aeronautics," *N.A.C.A. Report* No. 116 [1921], by permission of the National Advisory Committee for Aeronautics.)

coefficient and inversely proportional to the aspect ratio, at least for large aspect ratios. If there were only induced drag, the drag would be zero when the lift is zero. However, as shown in Fig. 28, this is not the case because there exist also wake drag and frictional drag. These two we cannot separate by simple measurement; their sum is, as noted above, the profile drag. If we assume that the profile drag is independent of the aspect ratio, then by using Prandtl's theory we can reduce the polar diagram for a certain aspect ratio to that for another aspect ratio. This was actually done in Fig. 29, where the measured values for wings of various aspect ratios are reduced to the curve for aspect ratio 5 by means of Prandtl's theory. The result indicates that the theoretical prediction is very nearly correct. A certain scattering occurs, but only for the wing of aspect ratio 1, where the basic assumption of the lifting-line theory does not hold.

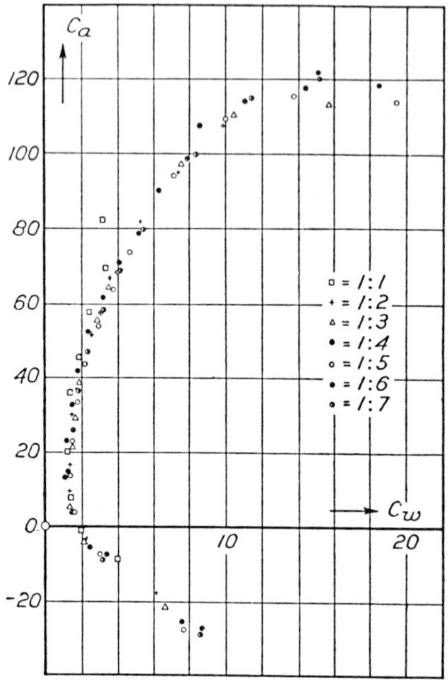

Fig. 29. Same data as Fig. 28, corrected to aspect ratio 5 according to Prandtl's theory. (From L. Prandtl, "Application of Modern Hydrodynamics to Aeronautics," *N.A.C.A. Report* No. 116 [1921], by permission of the National Advisory Committee for Aeronautics.)

I have described induced drag as the drag which must be counteracted in order to obtain lift. Thus we arrived at the concept of induced drag from the consideration that work must be done to create the downwash velocity related to the lift. Another explanation of induced drag which follows more closely the local flow phenomena is the following: Suppose that the airplane is flying in a horizontal direction. The wing is of finite span, so that free vortices spring from it and produce a field of induced velocities. The velocity induced at the wing itself is directed essentially downward; hence, when it is combined with the undisturbed relative velocity, the air appears to approach the wing along a slightly downward slope (Fig. 30). This means a reduction of the effective angle of attack and is responsible for the decrease of the slope of the lift curve mentioned in Chapter II. Then since the lift force caused by circulation is always perpendicular to the direction of the relative flow, it is inclined slightly backward from the perpendicular to the direction of flight. The component parallel to the flight direction is the induced drag. This explanation clarifies the fact that induced drag originates in pressure forces acting on the wing.

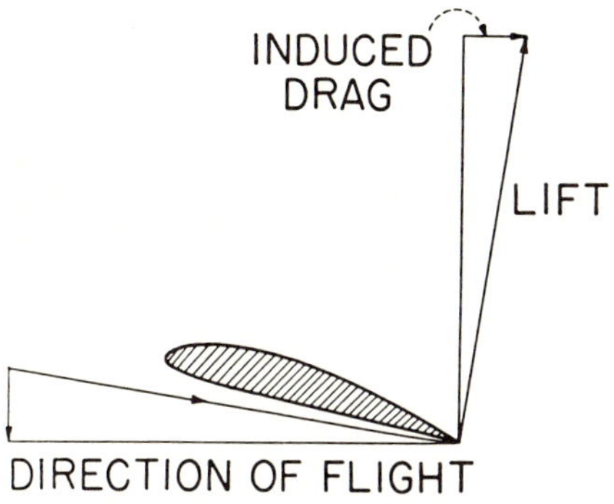

Fig. 30. Explanation of the production of induced drag.

I should like to make one more remark about induced drag. Induced drag is inevitable if there is lift and the wing span is not infinite. The question is how to make induced drag as small as possible. This problem was solved by Max Munk, a student of Prandtl (Ref. 2), in his doctor's thesis at Göttingen. Munk later came to this country to work for the National Advisory Committee for Aeronautics, became a professor at the Catholic University of America, and has also been associated with the work of the Naval Ordnance Laboratory. He showed that the minimum induced drag is obtained if the distribution of lift over the span corresponds to an ellipse. Such a distribution of lift is called an *elliptic distribution.*

Getting this result involved a lot of work. I remember when

Prandtl was working on his lifting-line theory in the summer of 1914 and I, having a commission in the Austro-Hungarian Army, was called home and passed through Göttingen.

"Now look here," Prandtl told me, "I am calculating these damned vortices and can't get a reasonable result for the induced drag. I tried to make the lift suddenly drop to zero at the wing tips, but the induced velocity becomes infinite. All right, I thought, Nature does not like such a discontinuity, so I made the lift increase linearly with the distance from the wing tip. That did not work either. This distribution of lift also does not produce finite induced velocity at the tip."

"Well, this is interesting. I will think it over, too," said I.

But I was too busy with the war to study the problem. Prandtl continued to work on it and later found the solution. It is, more or less, a mathematical trick: the problem can be solved if the lift is assumed to start with the one-half power of the distance from the wing tip, as, for example, in the case of the elliptic distribution found by Munk. Munk was one of Prandtl's most important collaborators during this period, and his contribution was certainly a significant part of the whole picture of wing theory.

According to the Prandtl-Munk formula, the minimum induced drag of a wing which produces a lift, L, is equal to $2L^2/\pi\rho U^2 b^2$, where b is the span of the wing, U is the velocity of flight, and ρ is the density of the air. Therefore the minimum power, P, required for sustentation of a weight, W, is given by

$$P = \frac{2W^2}{\pi\rho U b^2}.$$

It is interesting to compare this result with the early speculation of Renard, which we sketched in Chapter I. His formula for the power required for sustentation was

$$P = \frac{W^2}{2\rho U S},$$

where S is the wing area. This formula is based on an empirical law for the normal pressure exerted on a flat plate. The two formulas, ancient and modern, coincide if we take the aspect ratio equal to $4/\pi$ or 1.27. The ancient theory was rather pessimistic, because the power required is considerably reduced at the larger aspect ratios that are used in most airplanes.

Although for aerodynamic performance, especially for favorable lift-drag ratio and long range, very large aspect ratios would be desirable, structural considerations limit the practical values for moderate-speed airplanes to around 8 or 10. A significant exception is the transport airplane recently built by Hurel-Dubois in France, which has an aspect ratio of about 20. Assertedly a specifically designed strut inserted between fuselage and wing secures the necessary structural rigidity of the wing without excessive penalty in weight. For airplanes approaching or surpassing sonic velocity, the induced drag is relatively small in comparison with the other drag components; hence, in such airplanes, the designers usually employ small aspect ratios, down to 2 or even 1.5.

Wake Drag and Vortex Street

Now we come to the question of wake drag. According to D'Alembert's theory the wake drag is zero. Kirchhoff and Rayleigh tried to avoid this conclusion by assuming that surfaces of discontinuity are formed at the edges of the plate (see Chapter I). Physically, however, this is quite improbable, because it means that an infinite mass of fluid is carried with the plate as "dead fluid." This remains improbable even when the plate is accelerated from rest very slowly. It must be admitted that the real flow pattern is not well understood. Take, for example, the apparently simple problem of a sphere moving uniformly in a fluid; we do not know exactly how the flow pattern looks.

There is, however, at least one case in which we know something more definite about the flow pattern: this is the flow around an infinitely long cylinder. Fig 31 is a picture taken by a stationary camera of a circular cylinder moving to the left through a fluid originally at rest. We observe a double row of alternating vortices following the cylinder. The vortices in the upper row are turning clockwise, while those in the lower row are turning counterclockwise. This system of vortices replaces the infinite mass of fluid assumed to follow the body in the theory of Kirchhoff and Rayleigh. Indeed, the surfaces of discontinuity assumed in this theory can be considered as vortex sheets, and one finds, in general, that such vortex sheets are unstable. Also they have the tendency to roll up so that the vorticity concentrates around certain points.

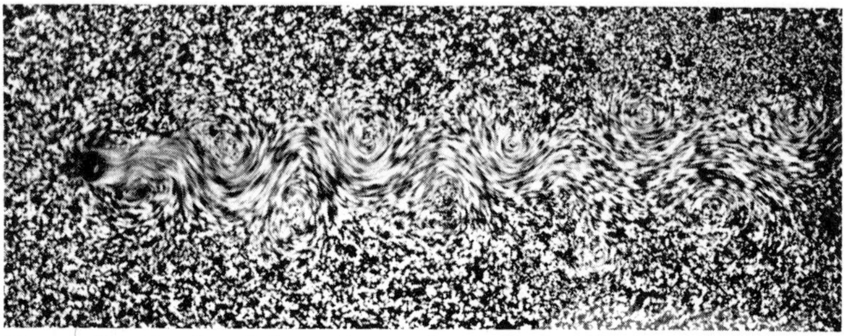

Fig. 31. Double row of alternating vortices behind a circular cylinder. (From I. Tani, *Fluid Mechanics* [in Japanese], [copyright 1951, Iwanami Shoten], by permission.

The arrangement of the vortices shown in Fig. 31 is connected with my name; it is usually called a *Kármán vortex street* or a *Kármán vortex trail*. But I do not claim to have discovered these vortices; they were known long before I was born. The earliest picture in which I have seen them is one in a church in Bologna, Italy, where St. Christopher is shown carrying the child Jesus across a flowing stream. Behind the saint's naked foot the painter indicated alternating vortices. Alternating vortices behind obstacles were observed and photographed by an English scientist, Henry Reginald Arnulpht Mallock (1851–1933) (Ref. 3), and then by a French professor, Henri Bénard (1874–1939) (Ref. 4). Bénard did a great deal of work on the problem before I did, but he chiefly observed the vortices in very viscous fluids or in colloidal solutions and considered them more from the point of view of experimental physics than aerodynamics. Nevertheless, he was somewhat jealous because the vortex system was connected with my name, and several times—for example, at the International Congresses for Applied Mechanics held in Zurich (1926) and in Stockholm (1930)—claimed priority for earlier observation of the phenomenon. In reply I once said, "I agree that what in Berlin and London is called 'Kármán Street' in Paris shall be called 'Avenue de Henri Bénard.' " After this wisecrack we made peace and became quite good friends.

What I really contributed to the aerodynamic knowledge of the observed phenomenon is twofold (Ref. 5): I think I was the first to show that the symmetric arrangement of vortices (Fig. 32, *upper*), which would be an obvious possibility to replace the

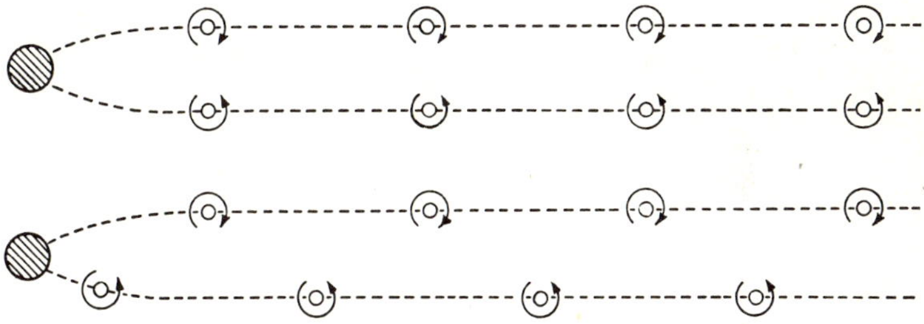

Fig. 32. Double rows of alternating vortices; symmetric (*upper*) and asymmetric (*lower*) arrangements.

vortex sheet, is unstable. I found that only the asymmetric arrangement (Fig. 32, *lower*) could be stable, and only for a certain ratio of the distance between the rows and the distance between two consecutive vortices of each row. Also, I connected the momentum carried by the vortex system with the drag and showed how the creation of such a vortex system can represent the mechanism of the wake drag—a point for which neither Mallock nor Bénard cared very much.

Perhaps I should tell how I became interested in the problem. In 1911 I was a graduate assistant in Göttingen. At that time Prandtl's main interest was in the theory of the boundary layer (which we will take up later), i.e., the flow of the fluid very close to the surface of a body. Prandtl had a doctoral candidate, Karl Hiemenz (Ref. 6), to whom he gave the task of constructing a water channel in which he could observe the separation of the flow behind a cylinder. The object was to check experimentally the separation point calculated by means of the boundary-layer theory. For this purpose, it was first necessary to know the pressure distribution around the cylinder in a steady flow. Much to his surprise, Hiemenz found that the flow in his channel oscillated violently.

When he reported this to Prandtl, the latter told him: "Obviously your cylinder is not circular."

However, even after very careful machining of the cylinder, the flow continued to oscillate. Then Hiemenz was told that possibly the channel was not symmetric, and he started to adjust it.

I was not concerned with this problem, but every morning when I came in the laboratory I asked him, "Herr Hiemenz, is the flow steady now?"

He answered very sadly, "It always oscillates." [1]

Now, I thought, if the flow always oscillates, this phenomenon must have a natural and intrinsic reason. One weekend I tried to calculate the stability of the system of vortices, and I did it in a very primitive way. I assumed that only one vortex was free to move, while all the other vortices were fixed, and calculated what would happen if this vortex were displaced slightly. The result I got was that, provided a symmetric arrangement was assumed, the vortex always went off from its original position. I obtained the same result for asymmetric arrangements but found that, for a definite ratio of the distances between the rows and between two consecutive vortices, the vortex remained in the immediate neighborhood of its original position, describing a kind of small closed circular path around it.

[1] Hiemenz later succeeded in making the flow almost steady by introducing still water into the wake region from below.

I finished my work over the weekend and asked Prandtl on Monday, "What do you think about this?"

"You have something," he answered. "Write it up and I will present your paper in the Academy."

This was my first paper on the subject. Then because I thought my assumption was somewhat too arbitrary, I considered a system in which all vortices were movable. This required a little more complicated mathematical calculation, but after a few weeks I finished the calculation and wrote a second paper.

Some people asked, "Why did you publish two papers in three weeks? One of them must be wrong." Not exactly wrong, but I first gave a crude approximation and afterward refined it. The result was essentially the same; only the numerical value of the critical ratio was different.

Now these vortices have many physical applications. Shortly after the publication of my paper, Rayleigh (Ref. 7) got the idea that the alternating vortices must give the explanation of the Aeolian harp—the singing wires. Some people will still remember the singing wires of the biplane cellules. The singing comes from the periodical shedding of vortices. When certain struts used on an underwater vehicle sang a high tune, Gongwer (Ref. 8) showed experimentally that the vibration was caused by the periodical shedding of vortices, which occurred when the trailing edges were not properly sharp. This also explains the singing of marine propellers, as was previously found by Gutsche (Ref. 9).

A French naval engineer told me of a case where the periscope of a submarine was completely useless at speeds over 7 knots under water, because the rod of the periscope produced periodic vortices whose frequency at a certain speed was in resonance with the natural vibration of the rod. Radio towers have shown resonant oscillations in natural wind. The galloping motion of power lines also has some connection with the shedding of vortices. The collapse of the bridge over the Tacoma Narrows was also caused by resonance due to periodic vortices. The designer wanted to build an inexpensive structure and used flat plates as side walls instead of trusses. Unfortunately, these gave rise to shedding vortices, and the bridge started torsional oscillations, which developed amplitudes up to 40° before it broke. The phenomenon was a combination of flutter and resonance with vortex shedding. I am always prepared to be held responsible for some other mischief that the Kármán vortices have caused!

I want to mention briefly the problem of reducing wake drag. As I explained at the beginning of this chapter, wake drag is caused by the fact that the streamlines do not follow the entire surface of the body but separate from it at some point. For example, in a circular cylinder, the streamlines separate from the surface somewhere midway between the front and rear, thus leaving downstream an eddying region of considerable extent. Such a separation may be, if not completely eliminated, at least postponed if we carefully shape the contour of the body, especially at the rear, so that the streamlines can follow the surface as far as possible. An airship hull is a good example of this sort of body, which is usually called a *streamline body.* Another example is the slender wing profile shown in Fig. 53 (p. 126).

For such a streamline body the paradox of D'Alembert is very nearly correct, since the pressures acting on the front and rear parts of the body are almost in balance. The body still experiences a drag, because there are frictional forces acting on the surface and also because the pressure forces cannot be completely balanced. However, the drag is usually rather small. For example, the drag of a carefully shaped airship model can be reduced to a value less than one-fiftieth of that of a disk of the same diameter placed normal to the stream. The distribution of pressure acting on such a body, except in the area near the rear end, can be calculated by the theory of nonviscous fluids with sufficient accuracy. The so-called method of sources and sinks first suggested by W. J. Macquorn Rankine (1820–1872) has been used for this purpose by a number of investigators. One practical method of solution actually used in the design of Zeppelin airships was put forward in a paper I published in 1927 (Ref. 10).

In airplane design the principle of "streamlining" has been extensively applied to achieve reduction of drag, for example, by retracting landing gears, attaching fillets to the junction between wing and fuselage, fairing the lines of the cockpit and windshield, and the like. Thus the wake drag of a modern high-speed airplane with clean lines has been reduced to a very small value. What remains to be done is to reduce the residual part of the drag, i.e., the skin friction. This problem will be discussed toward the end of this chapter.

Reynolds Number

If we investigate the drag phenomenon further, we see that the case in which drag is produced by vortex shedding is a special one. For example, if we measure the drag of a circular cylinder moving at various velocities, we find three different regimes of velocity. When the velocity is sufficiently small, the drag is proportional to the velocity; hence the drag coefficient is inversely proportional to the velocity. No alternating vortices can be observed in this speed regime. When the velocity is increased, the drag coefficient becomes almost independent of the velocity, and we can observe a regular pattern of alternating vortices. When the velocity is further increased, the periodical vortex shedding still persists, but the beautiful regular pattern no longer exists. Then, more or less suddenly, the drag coefficient drops to a substantially smaller value.

Now the question arises, What determines these curious changes in the magnitude of the same coefficient? This question is related to a fundamental problem that was first studied in 1883 by Osborne Reynolds (1842–1912), a professor at the University of Manchester (Ref. 11). The problem is, What is the prevailing law of similarity in fluid mechanics?

Before discussing this problem, however, I should explain something about the nature of fluid friction. Fluid friction is not like solid friction, such as the friction between a book and a desk when the book as a whole is made to slide over the surface of the desk. The action of friction between moving fluid and a solid surface is better illustrated by the following example: Suppose that a book containing many pages is placed on a desk and the upper cover is slowly pushed parallel to the surface of the desk. The pages slide over each other, but the lower cover sticks to the desk. Similarly, fluid particles stick to the surface of a body, so that there is no slip between fluid and solid surface. Near the surface, however, the fluid velocity increases with the distance from the surface, i.e., it exhibits a certain gradient. The velocity gradient across the flow produces friction between successive fluid layers which we call viscous friction. The sticking of the fluid to the surface is probably explained by the molecular or atomic structures of the solid and the fluid. Both consist of particles, atoms or molecules. The motion of the molecules in an airstream consists of a forward motion in the stream direction, on which a random motion is superposed. The atoms of the solid have a fixed mean position with empty spaces between. In general, according to the physicists, there is much more empty space in the world than space occupied by matter. If the molecules enter the empty spaces of the solid, they lose their forward velocity by collision with the solid molecules; and, if they rebound, they return with random velocity without preference for any flow direction. Hence the average velocity of the airflow right at the surface is zero, or equal to the velocity of the solid when the solid is moving.

At very high altitudes where the density of the air is very small and the air molecules are very far apart, the air can slip at the solid surface, as one solid slips on another one. (The branch of aerodynamics dealing with such phenomena is called *superaerodynamics*, but we shall forget about it for the present and ignore the flow of such low-density air.) We assume, therefore, that the velocity of the air is identical with the velocity of the solid at the surface and that the friction acting both at the surface and in the interior of the flow is viscous friction, determined by the gradient of the velocity across the flow.

The law governing viscous friction was originally given by Newton (Ref. 12) and later generalized in the form of a system of mathematical equations by Claude Louis M. H. Navier (1785–1836) (Ref. 13) and Sir George G. Stokes (1819–1903) (Ref. 14). It is assumed that the tangential force acting on a unit area between two adjacent layers of fluid is proportional to the gradient of the velocity across the flow. The constant of proportionality is called the coefficient of internal friction or viscosity, and is one of the characteristic physical constants of the fluid. It is large for "sticky" fluids like lubricating oil and small for "watery" fluids like water itself or air.

Let us now consider flow phenomena in which the geometrical arrangements as to the shapes of the boundaries or immersed bodies are similar. For example, we consider two flow patterns in each of which a sphere moves with a uniform velocity through an infinitely extended fluid at rest. The diameter of the sphere, the velocity of motion, and the density and viscosity of the fluid may be different. We want to find the condition that will permit the flow pattern to remain similar. In other words, we want to find the law of mechanical similarity for geometrically similar arrangements.

First, all the forces acting on a fluid element must be listed. These are gravity, friction, force of inertia, and pressure. Let us forget for a while about gravity, since gravity usually has no noticeable influence in aerodynamic phenomena which have local character, although it is important in large-scale phenomena like those treated in the science of weather. In an incompressible fluid, the pressure is a kind of passive reaction, whose magnitude is just sufficient to balance the other forces acting on a fluid element. Hence it is enough for us to consider the friction and the inertia forces. If the ratio between these two forces remains unchanged, the flow pattern will remain similar.

The inertia force acting on a fluid element is equal to the rate of change of the momentum in unit time. The length scale of the pattern may be characterized by an appropriately chosen length L, e.g., the diameter of the sphere. If U is the characteristic velocity, such as the velocity of motion, the time scale of the phenomenon is given by L/U. Finally, let ρ be the density and μ the coefficient of viscosity of the fluid. Then the masses of the two corresponding fluid elements in the two flow patterns will be in the ratio ρL^3, the respective values of the momentum in the ratio $\rho L^3 U$, and the rates of change of the momentum in the ratio $\rho L^3 U \times U/L$, or $\rho L^2 U^2$. We could have started with this expression by arguing that the inertia force must be proportional to the dynamic pressure $\frac{1}{2}\rho U^2$ and that the force acting on corresponding elements is therefore proportional to $\frac{1}{2}\rho U^2 \times L^2$.

The frictional force acting on a unit area is proportional to $\mu U/L$, because it is equal to the velocity gradient across the flow multiplied by the coefficient of internal friction μ. The resultant of the frictional forces on a fluid element is then proportional to $(\mu U/L) \times L^2$, or μUL. Hence the ratio between the inertia and the frictional forces is proportional to

$$\frac{\rho U^2 L^2}{\mu UL} = \frac{\rho UL}{\mu} \quad \text{or} \quad \frac{UL}{\nu},$$

where $\nu = \mu/\rho$ is called the coefficient of kinematic viscosity. If we compare stresses, i.e., forces acting on a unit area of a fluid element, we find that, in order to achieve mechanical similarity, the normal stress or pressure proportional to ρU^2 must be in a constant ratio to the tangential or frictional stress, proportional to $\mu U/L$.

In summary we may say that, if the ratio UL/ν has the same numerical value for two flow systems, then we may expect the flow patterns to remain similar. In other words, if the diameter of the sphere of the first system is twice as large as that of the second system, then we must make the velocity of the sphere of the first system equal to half of the velocity in the second system in order to get similar flow patterns, provided the motion takes place in a fluid with the same kinematic viscosity. If the kinematic viscosity of one system is one-tenth that of another, the product of the linear dimension and the velocity of the first system must also be ten times smaller in order for the flow patterns of the two systems to be similar. The expression UL/ν is a nondimensional quantity and is called the *Reynolds number*.

One very illustrative example of the similarity law announced in the past paragraph is the method of increasing the Reynolds number in wind-tunnel experiments. In general, the dimensions of a wind-tunnel model are reduced in a certain scale relative to the prototype. Nevertheless, mechanical similarity can be achieved by using a fluid of low kinematic viscosity, an idea independently suggested by Margoulis (Ref. 15) and Munk (Ref. 16). Munk, in particular, considered the feasibility of a wind tunnel utilizing compressed air as its working fluid, and following this idea the *variable-density wind tunnel* was constructed at the N.A.C.A. Since the coefficient of viscosity, μ, of a gas is but little influenced by the density or pressure, the effect of increasing the pressure is to reduce the kinematic viscosity.

According to the kinetic theory of gases, the coefficient of internal friction, μ, is proportional to $\rho c \lambda$ where c is the mean molecular velocity of thermal agitation and λ is the mean free path of a molecule. Thus, except for a numerical factor, the Reynolds number may also be expressed by

$$\frac{U}{c} \times \frac{L}{\lambda},$$

that is, the product of the ratio of the velocity of the body to the molecular velocity and the ratio of the linear dimension of the body to the mean free path. As we shall show in the next chapter, the molecular velocity is of the same order of magnitude as the sound velocity; the mean free path for air at normal conditions is very small—of the order of two-millionths of an inch in length. Hence as far as ordinary low-speed flow is concerned, U/c is small and L/λ is very large, and only their product appears in the law of similarity in the form of the Reynolds number. However, if the velocity is near the velocity of sound, the ratio U/c is no longer small, and it appears separately as a second similarity parameter. This parameter is called the *Mach number*, and we shall encounter it as the governing parameter in the next chapter.

Under certain conditions the ratio L/λ becomes a second, independent, governing parameter, namely, if the mean free path is comparable to the dimension of the body. This occurs, for example, in the case, already mentioned, of a body moving in air of very low density, e.g., at high altitude. In this case we are in a range where the mechanics of continuous fluids no longer apply, and collisions between molecules have to be considered.

Let us exclude the cases where the velocity is commensurable with sound velocity and the mean free path is commensurable with the body dimensions. Then the Reynolds number is the sole governing parameter, and if the Reynolds number has the same value the flow is similar and therefore all nondimensional coefficients must have the same values. In other words, the nondimensional parameters are in general to be considered as functions of the Reynolds number. I was not quite correct when I said that the drag coefficient of a circular cylinder depends only on the velocity. This is true if the diameter of the cylinder and the kinematic viscosity of the fluid are kept unchanged. In fact, the drag coefficient of a circular cylinder depends on the Reynolds number, as shown in Fig. 33.

It is an interesting fact that neither Reynolds himself nor other British scientists who followed him gave a specific name to the nondimensional parameter UL/ν; it was Arnold Sommerfeld (1868–1952) who named the parameter in honor of Reynolds in 1908. The Reynolds number is now generally used in hydrodynamics, aerodynamics, hydraulics, and other sciences which have to do with fluid flow. It works in some cases almost like black magic.

I recall the following experience: In 1911 a well-known Ger-

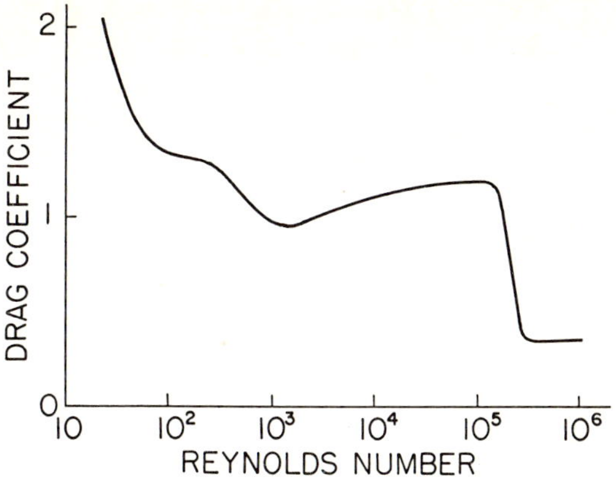

Fig. 33. The drag coefficient of a circular cylinder as a function of the Reynolds number.

man physical chemist, Emil Bose, published a paper containing very careful measurements on the pressure drop in pipe flow of various organic liquids (Ref. 17). He used identical apparatus for all liquids and measured the time required for equal volumes of different liquids to flow through the same pipe, and the corresponding pressure difference between the two ends of the pipe. Comparing the results for different liquids, he found that chloroform, for example, is less viscous than water at low speeds, but it behaves almost the same as water at higher speeds; bromoform is more viscous than mercury at low speeds, but it becomes "less viscous" than mercury at higher speeds. Apparently, "less viscous" in this case means that a smaller pressure difference is required for the same rate of flow. I suggested the use of the Reynolds number—defined as the mean velocity multiplied by the pipe diameter and divided by the kinematic viscosity—as parameter and found that the formulas which Bose proposed for the representation of his experimental results with nine liquids could be unified into one single formula (Ref. 18).

Fig. 34 represents in logarithmic scale the measured pressure drop, P, versus the time required, T, for the outflow of the same

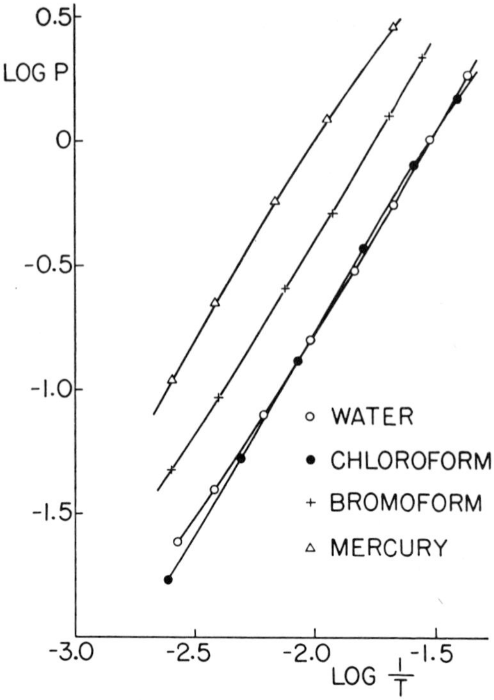

Fig. 34. Experimental results of Bose on pressure drop in pipe flow. The logarithm of the pressure drop P in kilograms per square centimeter is plotted against the logarithm of $1/T$, where T is the time in seconds required for outflow of the same volume (8.81 cubic centimeters) of liquid. (From data of E. Bose, D. Rauert, and M. Bose, in *Physikalische Zeitschrift*, *10* [1909], 406–409, and *12* [1911], 126–135.)

volume, Q, of four selected liquids ($Q = 8.81$ cubic centimeters). In Fig. 35 the nondimensional quantity, PT/μ, is plotted as a function of another nondimensional quantity, $\rho Q^{2/3}/\mu T$, which for similar geometrical arrangement of the apparatus is proportional to the Reynolds number. It is seen that the data shown in the four curves of Fig. 34 all lie on a single curve. This proves not so much the correctness of the similarity law, which does not need experimental proof, but the exactitude of Bose's measurements.

With apology to the hydraulicists who may read this book, I confess that I used to call hydraulics "the science of variable constants." The truth is that most constants appearing in the old hydraulics books are simply functions of the Reynolds number. After the concept of the Reynolds number was adopted by the hydraulicists and the chemical engineers, the whole subject of flow in pipes and channels became much clearer. However, it was a long time before the full importance of Reynolds' ideas penetrated the minds of physicists, chemists, and engineers. In American hydraulic literature of the twenties an equivalent of the Reynolds number appears as the "turbulence factor."

I said that the Reynolds number works like black magic, because in engineering one can sometimes use a similarity rule and other general methods for the reduction of parameters without much of an understanding of the phenomena.

This reminds me that a great engineer, Charles F. ("Boss") Kettering, then director of research for General Motors, once told me when I had lunch with him and the late Robert A. Millikan,

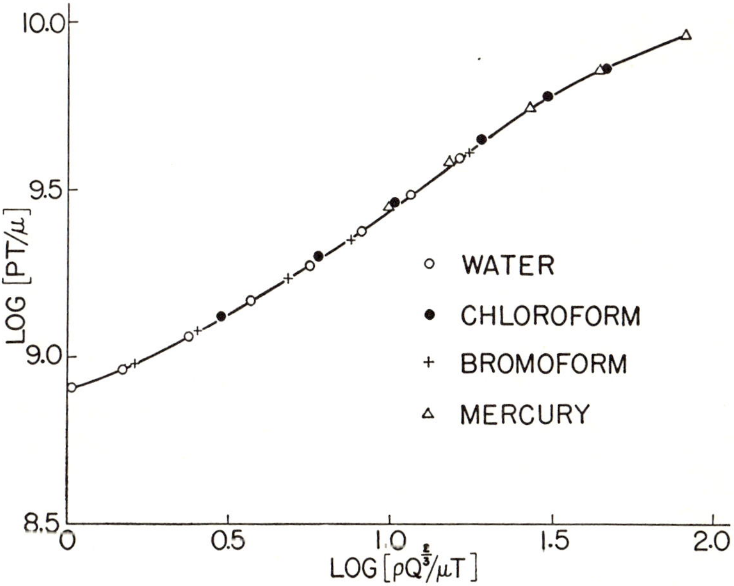

Fig. 35. Nondimensional representation of Bose's experimental results plotted in Fig. 34. Same notation as in Fig. 34; ρ and μ are density and viscosity of liquids, respectively. (From data of Th. von Kármán, in *Physikalische Zeitschrift*, *12* [1911], 283–284.)

"I must confess that thermodynamics was always black magic for me!"

This is an interesting observation by a great practical engineer, who certainly had to apply the entropy law and other thermodynamic rules in his work!

Laminar and Turbulent Flow

One curious feature that can be seen in Fig. 33 is the sudden decrease in the drag coefficient of the circular cylinder in the neighborhood of the Reynolds number 2×10^5. This phenomenon of sudden change in the value of the drag is not confined to circular cylinders but occurs also with spheres and other bodies; it is characteristic for many other phenomena in fluid mechanics. The physical reason for such a sudden change is the existence of two fundamentally different types of flow, which we call *laminar* and *turbulent* flows.

In 1883 Reynolds carried out a series of experiments on flow in tubes. One of his experiments is shown diagrammatically in Fig. 36. A long glass tube was connected to a reservoir, and the flow through the tube was observed by introducing a dye at the entrance of the tube. At small velocities the dye forms a thin,

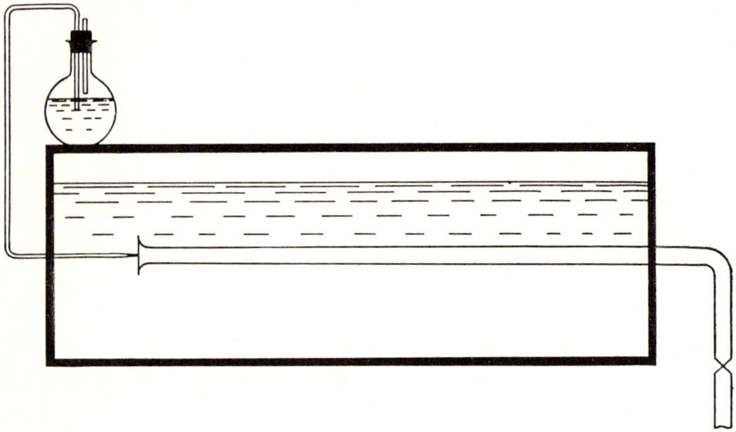

Fig. 36. Diagrammatic sketch of Reynolds' experiment.

straight thread parallel to the axis of the tube, indicating that flow is steady and orderly in nature. This type of flow we call a laminar flow. If the velocity is increased by small steps, one observes at a certain velocity a sudden change in the character of the flow; the thread becomes violently agitated, and the dye quickly spreads over the whole tube. The flow changes from the laminar type to one of an oscillatory or, rather, irregular character, which we call turbulent flow. Turbulent flow is much more common in nature and in engineering devices than laminar flow. For example, the flow of water in rivers and the motion of the air in the atmosphere are practically always turbulent. The fluid motions with which the engineer is concerned are turbulent in most cases.

To be sure, Reynolds was not the first to observe and analyze the phenomenon of turbulent flow. Indeed, the German engineer, Gotthilf Heinrich Ludwig Hagen (1797–1884) (Ref. 19), recognized the transition from laminar to turbulent flow in 1854. Reynolds, however, carried out a systematic series of experiments and demonstrated that the transition from laminar to turbulent flow occurs when the parameter we call the Reynolds number exceeds a certain critical value. The Reynolds number in this case may be defined by taking the diameter of the tube and the mean velocity over the cross section of the tube as the characteristic length and velocity, respectively.

Now, the characteristic feature of turbulent flow is that it is quite irregular. However, genuinely regular motion is exceptional in nature. Even laminar flow appears regular only to the human observer who looks at the molecular motion from so far away that he can see only the average motion. Similarly, the velocity that the practical engineer measures in the turbulent flow of a river is actually the mean value of a velocity component, because his measuring instrument is not sufficiently refined to follow the irregular motion. If he had finer measuring instruments, he could observe the instantaneous values of the velocity. The presence of the irregular motion radically changes the flow pattern, especially

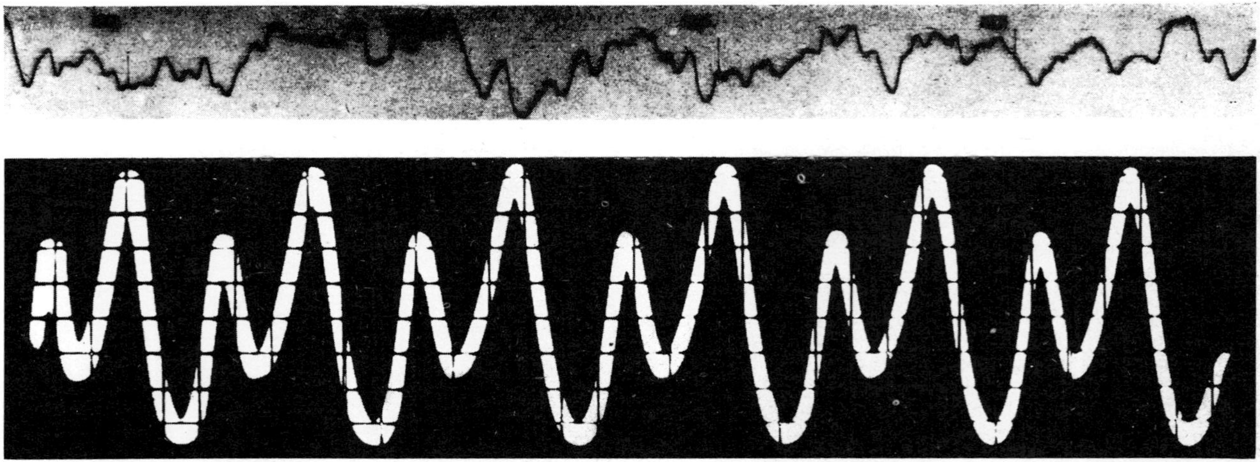

Fig. 37. Oscillograph records of velocity fluctuations in turbulent flow (*upper*) and alternating vortex street (*lower*). (Courtesy Guggenheim Aeronautics Laboratory, California Institute of Technology.)

the energy losses. But the irregular motion is so complicated that it is generally hopeless to follow all individual details of the flow. Moreover, what we are interested in for practical purposes are mostly average quantities. The real mechanism of turbulent motion must be treated by statistical methods.

Fluid motion can be analyzed from two different points of view: the so-called *Eulerian method* considers pressure and velocity at a fixed point, and the *Lagrangian method* describes the fate of an individual particle.

From the viewpoint of the Eulerian method, turbulent flow is described by the fluctuations of velocity and pressure at a given point. Fig. 37 shows oscillograph records of velocity as a function of the time. The records were made by a *hot-wire* anemometer, in which a platinum wire of very small diameter is exposed to the flow at a fixed point and is heated electrically. When the velocity of flow changes, the temperature of the wire and therefore its electric resistance change; this change can be recorded by suitable instruments. The development of the hot-wire technique is due to several experimental aerodynamicists; recently Hugh L. Dryden and his collaborators at the National Bureau of Standards made significant contributions (Ref. 20). The upper record of Fig. 37 represents a typical case of turbulent flow. The lower record is obtained by placing the wire in an alternating vortex street, such as described in a preceding section of this chapter. We clearly observe the very irregular character of the turbulent fluctuation, which contains all possible frequencies, while one definite frequency prevails in the vortex street. The difference between the flow due to vortex shedding and the turbulent motion can be illustrated by a column of soldiers marching in step and a crowd of people moving in a haphazard fashion.

If we look at the turbulent flow from the Lagrangian viewpoint —for example, by adding small particles that will move with the fluid, thus making the flow visible—we observe a continuous intermingling of particles instead of fluctuation at a fixed point.

In Reynolds' experiment described above, we assume that the dye particles are carried by the fluid elements. This is why the dye spreads over the whole tube when the flow changes from laminar to turbulent. The turbulent intermingling of fluid particles also changes the velocity distribution in the tube in such a way that the velocity differences in the central part of the tube are reduced, and thus the distribution is more nearly uniform when the flow is turbulent than when it is laminar.

Fig. 38 shows the velocity distributions for the two types of flow, as given by measurement and drawn for the same amount of fluid flowing per second. Since the velocity near the center is more uniform, the velocity gradient at the wall must be considerably greater when the flow is turbulent. Consequently, the friction loss in a turbulent flow is much greater than it is in a laminar flow carrying the same quantity of fluid.

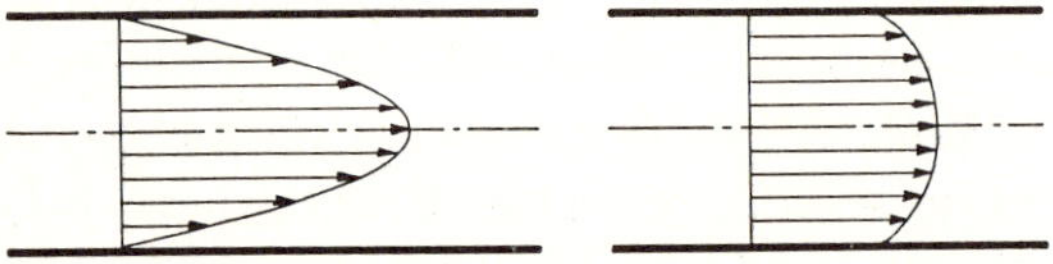

Fig. 38. Velocity distributions of flow in a tube, laminar (*left*) and turbulent (*right*).

Turbulence is not confined to the flow in tubes but also occurs, for example, in the flow just adjacent to the surface of a body moving in a fluid, the so-called *boundary layer*. The flow in this layer may be laminar at low Reynolds numbers and may become turbulent when the Reynolds number exceeds a certain critical value. This change has a favorable consequence because the violent intermingling of particles enables the turbulent layer to stick to the surface better than does the laminar layer, which contains less kinetic energy and leaves the surface earlier. At low Reynolds numbers, especially in the range where the drag coefficient of a sphere or cylinder is almost constant and has the larger value, the boundary layer is laminar and the early separation of the flow creates a broad wake filled by vortices. Then, at a certain higher Reynolds number, the flow in the boundary layer becomes turbulent, the separation is delayed, and the size of the wake is reduced. This explains the relatively sudden reduction of the drag coefficient at a certain Reynolds number (Fig. 33), mentioned above.

The phenomenon of the sudden change of sphere drag was first observed in a rather amusing way. Prandtl in Göttingen and Eiffel in Paris measured the drag of the sphere; Prandtl obtained a value for the drag coefficient which was more than twice that obtained by Eiffel.

They exchanged information, and one of the young engineers in Prandtl's laboratory said, "Oh, M. Eiffel forgot a factor of two. He calculated the coefficient referred to ρU^2, not $\frac{1}{2}\rho U^2$."

This remark somehow reached Paris and the elderly M. Eiffel became very angry. He then measured the drag for a wider range of Reynolds numbers—the maximum Reynolds number attainable was a little higher in his wind tunnel than Prandtl's—and discovered that a sudden decrease in the drag coefficient occurred beyond a certain Reynolds number (Ref. 21). Thus he discovered the dependence of the phenomenon on the Reynolds number.

But Eiffel did not find the physical reason for the sudden change. It was Prandtl (Ref. 22) who gave the explanation mentioned above. He also added an interesting experiment: a fine wire ring was put around a sphere a short distance in front of the separation point of the laminar layer. The wire disturbed the flow in the boundary layer, so that the transition to turbulence, and hence the sudden drop of drag, occurred at a smaller Reynolds number. Paradoxically, therefore, although the wire ring was an additional obstacle, the total drag was reduced by the presence of the wire because laminar separation was prevented.

Skin Friction and Boundary Layer

The problem of skin friction acting on flat plates moving parallel to their surface through a fluid was of primary interest for shipbuilders. In the years 1793–1798 Mark Beaufoy in England carried out systematic experiments on fluid resistance in general and on the magnitude of skin friction in particular. The results of his experiments were published by his son, Henry Beaufoy, in 1834 (Ref. 23). Many years later, in 1872, William Froude published the results of a series of important experiments on the subject. Froude's second report (Ref. 24), dated December 1872, is a remarkable document, because, I believe, it was the first time that an author clearly stated that the frictional force must have its counterpart in the loss of momentum of the fluid which has passed along the surface of the plate. This is the fundamental idea of every modern theory of skin friction. However, the theoretical analysis of the phenomenon based on the equations of motion of fluids started with Prandtl's paper presented to the Third International Congress of Mathematicians held in 1904 in Heidelberg. Prandtl showed in his paper (Ref. 25) that for a fluid of small viscosity, such as air or water, the viscosity will substantially affect the flow only in a thin layer adjacent to the surface. Outside this layer, viscosity can be neglected and the flow can be described to a high degree of accuracy by the mechanics of nonviscous fluids.

Prandtl called the thin layer near the wall affected by viscosity the "*Grenzschicht*"; the term *boundary layer* is used in English terminology. He showed that the small thickness of the boundary layer permits essential simplifications in the equations of motion of a viscous fluid, so that the problem of frictional drag becomes accessible to mathematical analysis. Thus from 1904 on the boundary-layer theory became an important part of fluid mechanics. Some German scientists propose to publish an anniversary volume "Fifty Years of Boundary Layer Theory" in this year of 1954.

Prandtl first obtained the solution for a flat plate exposed to a uniform parallel stream. He found that if it is assumed that the flow in the boundary layer is laminar, the thickness of the layer increases with the square root of the distance from the leading edge of the plate, and the friction acting per unit area decreases inversely proportional to the square root of the same distance. Summing up the frictional force over the flat plate, we can obtain the total skin friction.

In a uniform external flow, the similarity of the velocity distributions across all sections of the boundary layer permits the reduction of the problem to the solution of an ordinary differential equation, i.e., of a differential equation with one variable. If the flow outside the boundary layer is not uniform, as in the case of a wing section, the problem requires in general the solution of a partial differential equation—a differential equation with two or three variables.

In the past fifty years a great number of scientific publications have been devoted to the solution of boundary-layer equations and to the comparison of theory and experiments. I proposed a simplified method in 1921 in one of my papers (Ref. 26); I used an integral relation describing the development of the boundary layer as a whole, instead of trying to solve the partial differential equation. This method has been employed extensively by many authors. Its usefulness was first demonstrated by Karl Pohlhausen (Ref. 27).

Boundary-layer theory also enables us to calculate the point where the flow separates from the surface, because—as Prandtl pointed out—the separation of flow occurs mainly because the kinetic energy is dissipated by viscosity within the layer. As I have already mentioned, the wake drag is caused by flow separation. It is therefore important to predict the conditions in which separation will occur. Before the introduction of the boundary-layer theory into fluid mechanics, separation could be predicted only if the flow passed over a sharp edge. The boundary-layer theory opens the possibility of predicting the separation of flow in the case of a surface without sharp edges, at least in cases when the external flow is known and the flow in the boundary layer is laminar.

In practical application, however, complications arise due to the transition from laminar to turbulent flow. As we have seen earlier, the flow in the boundary layer can be laminar or turbulent, just as in pipes or other conduits. As we mentioned, the transition from laminar to turbulent flow in the boundary layer causes the decrease in the drag coefficient of blunt bodies like spheres and circular cylinders. In a similar way the flow in the boundary layer of wing sections (called airfoil sections or, briefly, airfoils) may also change from laminar to turbulent. We know that a turbulent boundary layer can better resist the tendency of separation than a laminar boundary layer; it sticks to the surface better. We also know that the stall of an airfoil, i.e., the reaching of maximum lift at a given flight speed, is caused by flow separation. Hence the transition from laminar to turbulent flow may be beneficial in permitting airfoils to reach higher lift, just as it proved to be beneficial in reducing the wake drag of blunt bodies. This phenomenon was discussed in a paper published in 1935 by Clark B. Millikan and me (Ref. 28). Fig. 39, which is taken from that paper, illustrates the "play" between the points of transition and separation. However, as far as skin friction is concerned, turbulence in the boundary layer always works against the designer in that it increases the magnitude of the friction.

Here we touch upon a problem that is one of the most important and most difficult subjects in modern fluid mechanics—the problem of the turbulent flow and of the turbulent boundary layer in particular. The real theory of the mechanism of turbulence is a very complicated problem of statistical mechanics. As in statistical mechanics in general, we are dealing with a disorderly or chaotic motion. It is almost hopeless to follow the fates of individual particles, but we may obtain results concerning statistical mean values.

Many scientists are working on statistical turbulence theories. Interesting results have been obtained concerning a simple type of turbulence, which is uniform and isotropic in space (i.e., the statistical mean values are independent of location and orientation in space). Unfortunately, this type of turbulence cannot

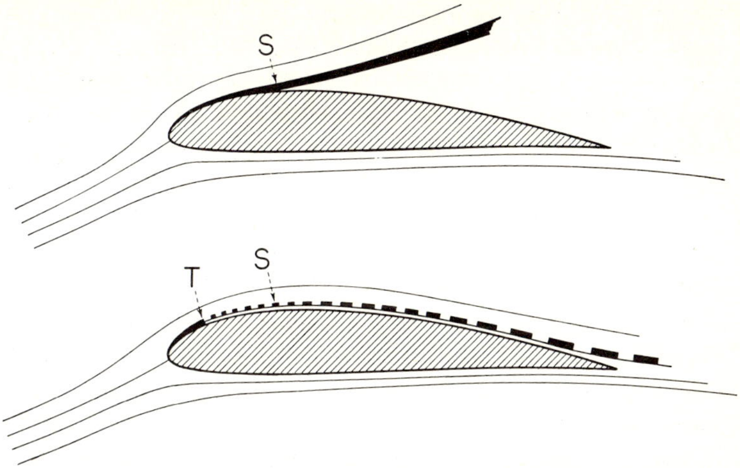

Fig. 39. Alternative flows around an airfoil section. When the Reynolds number is small, the transition point is downstream of the separation point, *S* (*upper*) and an early separation occurs. An increase in the Reynolds number causes the transition point, *T*, to be located upstream (*lower*); the boundary layer at the laminar separation point, *S*, is already turbulent and so clings to the airfoil surface. (From Th. von Kármán and C. B. Millikan, in *Journal of Applied Mechanics, 2* [1935], A–22, by permission of the American Society of Mechanical Engineers.)

transfer forces from fluid layer to fluid layer; therefore the statistical theory so far cannot be applied to turbulent friction. Nevertheless, the progress of the statistical theory is extremely promising, in spite of the difficulties from both the mathematical and physical points of view.

After the introduction of the fundamental concept of isotropic turbulence and initiation of its study by Sir Geoffrey I. Taylor of Cambridge University in 1935 (Ref. 29), Leslie Howarth and I made some significant advances (Ref. 30). Later, important new ideas were put forth independently by the Russian mathematician Andrei N. Kolmogoroff (Ref. 31) and the German physicist Werner Heisenberg (Ref. 32). I was in Moscow in 1945 and talked over turbulence problems with Kolmogoroff. He told me about the progress he had made in the statistical mechanics of turbulence; he published a paper on the subject in 1941, but it was not known until much later in Western Europe.

In the same year I went to London and repeated Kolmogoroff's story to my friend, Sir Geoffrey, who said, "That is exactly the same thing that Heisenberg tried to explain to me three months ago here in Cambridge!"

And we later found out—especially after George K. Batchelor investigated the two theories closely (Ref. 33)—that Kolmogoroff's reasoning and results were almost identical with those of Heisenberg. One man conceived the idea in Russia and the other one in Germany, both during the time in which the two countries were involved in a war for life or death. Heisenberg later gave a broader formulation to his theory, but the whole subject is still in flux. Among other scientists working on the problem are Chia-Chiao Lin of Cambridge, Massachusetts, and Subrahmanyan Chandrasekhar in Chicago.

It is important, however, to mention that useful semiempirical solutions for the computation of turbulent friction were found prior to the strictly statistical theory. To be sure, these semiempirical theories are also based on statistical concepts. Prandtl (Ref. 34) tried to transfer the concept of the mean free path used in the kinetic theory of gases into the theory of turbulence. In the kinetic theory of gases, the mean free path can be calculated, because the particles are well defined as molecules, whereas the fluid particles which intermingle in turbulent flow are somewhat ambiguous. However, Prandtl successfully introduced a certain path of convection or *mixing length* into a simplified picture of the turbulent mixing; in principle he left the magnitude of the mixing length to be determined by experiment.

I considered the problem from a somewhat more general point of view and introduced the assumption that the flow patterns of the turbulent flow in the neighborhoods of any two points in the flow are similar and differ only in their length and time scales (Ref. 35). It was then possible to correlate the mixing length with the velocity distribution by solving a specific differential equation. The velocity distribution calculated in this way agrees very well with measurements and is usually called the *logarithmic velocity distribution*, because the velocity is expressed by a logarithmic function of the distance from the surface. The same formula was obtained independently by Prandtl (Ref. 36), when he assumed that the mixing length is proportional to the distance from the surface.

There still remained the problem of establishing the connection between the fully developed turbulent region and the so-called *laminar sublayer*, which always exists next to a solid surface, where the surface prevents any turbulent fluctuation. Prior to the discovery of the logarithmic velocity distribution, several empirical distribution laws were tried, but one always had to change them when the range of experimental evidence was extended.

The formulation of the logarithmic law was the end result of a long struggle to obtain correlation between theoretical ideas and experimental evidence. Prandtl's school and my own worked on the problem in a spirit of co-operative rivalry. The logarithmic velocity distribution was first found for flow between two walls. But it could be applied without difficulty to the calculation of the skin friction of a flat plate which is covered by a turbulent boundary layer. Fig. 40 shows a plot of the nondimensional coefficient of skin friction as a function of the Reynolds number, which is referred to the length of the flat plate and the relative undisturbed velocity outside the boundary layer. The plot contains, in addition to the theoretically predicted values, a number of curves obtained experimentally during the past several decades. The agreement between theory and experiment is excellent, although it should be noted that one universal constant which is left open in the theory has been adjusted. In the same figure the coefficient of laminar friction is also shown, i.e., the friction coefficient of a flat plate covered by a laminar boundary layer. In the range of its validity the curve calculated by Prandtl's theory of laminar boundary layer agrees very well with experiment. C. B. Millikan and N. B. Moore have extended the laminar and turbulent theories, respectively, to slender bodies of revolution and calculated the skin friction for certain airship models (Ref. 37).

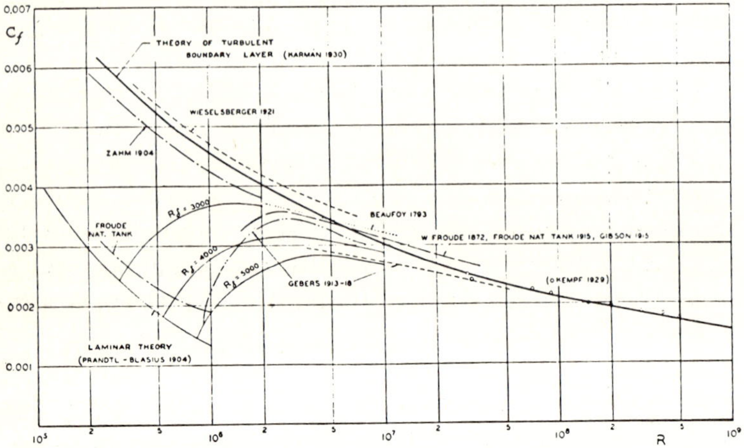

Fig. 40. The skin-friction coefficient C_f of smooth, flat plates as a function of the Reynolds number R. (From Th. von Kármán, in *Journal of the Aeronautical Sciences, 1* [1934], 13, by permission of the Institute of the Aeronautical Sciences.)

Fig. 40 includes some experimental curves that fit neither the laminar nor the turbulent curve. These correspond to cases where the boundary layer begins in a laminar state and becomes turbulent behind a certain point. We do not know enough about the mechanism of transition from laminar to turbulent flow to calculate theoretically how the transition occurs. What we can predict with some certainty is the condition in which a disturbance in the laminar boundary layer may increase with time. Small disturbances may either decay or grow with time; if they grow, we say that the laminar boundary layer is unstable.

The theory of instability of laminar flow, which has the aim of predicting the value of the Reynolds number at which disturbances no longer decay is a mathematical problem which has intrigued a number of prominent mathematicians. Sommerfeld

(Ref. 38) first attacked it; Heisenberg (Ref. 39) worked on it; and Tollmien (Ref. 40) and Lin (Ref. 41) finally completed the calculations. At first the validity of the theory was questioned, because there was no experimental evidence to support the theoretical predictions. But later it was made clear by Dryden, Schubauer, and Skramstad (Ref. 42) that the true phenomenon had been masked by the turbulence existing in the wind-tunnel stream. These investigators succeeded in creating a wind stream of extremely low turbulence where they could show that the predictions of the stability theory are correct and that the appearance of instability agrees with the commencement of the transition from laminar to turbulent flow.

The problem of the gradual development of transition is, however, much more complicated, and I think there is still much work to be done before we fully understand its mechanism. For example, almost all theories and experiments are concerned with a smooth body surface, while the surface of an actual airplane is more or less irregular and rough. When roughness exists on the surface, the disturbance caused by the roughness may cause a premature transition to turbulence. This problem was studied experimentally by Tani and Hama (Ref. 43) in Japan during the last war.

The influence of roughness also enters into the problem of turbulent skin friction. It appears that roughness has no significant effect on skin friction when the Reynolds number is below a certain limit. The physical reason is that below this Reynolds number the thickness of the laminar sublayer exceeds the height of the irregularities of the surface—called roughness elements—and these elements are not able to produce additional turbulence in the main stream. With increasing Reynolds number, the laminar sublayer becomes thinner and thinner, so that the roughness elements emerge and begin to influence the main flow. When the height of the roughness elements is large in comparison with the thickness of the laminar sublayer, the skin friction is apparently given by the total frontal drag of these elements. In this case every protuberance can be considered as a small blunt body, and its individual drag is proportional to the square of the velocity of the fluid stream which strikes it. This causes the coefficient of the total friction to depend only on the degree of roughness and to be independent of the Reynolds number of the plate.

As I mentioned earlier, turbulence works against the aeronautical engineer as far as skin friction is concerned. Hence the question arises whether there is any possibility of "cheating nature" and maintaining the boundary layer in a laminar state up to a higher-than-usual Reynolds number. In the period immediately before and during the last war, much attention was given to *laminar-flow airfoils*. These airfoils are so designed that the lowest pressure on the surface occurs as far back as possible. The reason for this design is the fact that the stability of the laminar boundary layer generally increases when the external flow is accelerated, i.e., in a flow with a pressure drop, while the stability decreases when the flow is directed against increasing pressure. Considerable reduction in skin friction is obtained by extending the laminar regime in this way, provided that the surface is sufficiently smooth.

I remember that, during our return voyage from the Volta Congress for High Speed Flight in 1935, Eastman N. Jacobs told me that in his opinion no really important progress originating from aerodynamic theory could any longer be expected. Jacobs is one of the most creative aerodynamicists in this country, and at that time he was working for the N.A.C.A. It is a remarkable coincidence that, a few years later, he himself most effectively contributed to the development of laminar-flow airfoils (Ref. 44). The success of the design of the airfoil was first announced by the late George W. Lewis, then director of research of the N.A.C.A., in his Wilbur Wright Memorial Lecture to the Royal Aeronautical Society in 1939, but details of the principle of the design were not given for reasons of national security (Ref. 45). The same problem was also pursued independently in England and in Japan, and, curiously enough, the principle of the design was first published in 1940 in a report of the Aeronautical Research Institute, Tokyo Imperial University (Ref. 46).

It is also possible to postpone the transition to turbulence by preventing the growth of the boundary layer beyond a certain limit by means of *boundary-layer control.* Restricting the thickness of the laminar boundary layer usually preserves the stability much longer than when the layer grows naturally. Boundary-layer control can be effected by removing air from the boundary layer through slots or holes in the wing surface, or through a porous wing surface. Such methods have been studied extensively at laboratory scale and also on a few flying models. The possibility exists that by application of the principles of boundary-layer control airplanes may be designed in the future with much lower drag than they now have. I do not know whether this will really be accomplished, but it would be a great victory for aerodynamic theory.

References

1. Prandtl, L., Betz, A., and Wieselsberger, C., *Ergebnisse der Aerodynamischen Versuchsanstalt zu Göttingen*, I (Munich and Berlin, 1923), 50–53.
2. Munk, M. M., *Isoperimetrische Probleme aus der Theorie des Fluges*, Göttingen Dissertation (1918).
3. Mallock, A., "On the Resistance of Air," *Proceedings of the Royal Society of London*, series A, *79* (1907), 262–265.
4. Bénard, H., "Formation de centres de giration à l'arrière d'un obstacle en mouvement," *Comptes rendus de l'Académie des Sciences, Paris*, *147* (1908), 839–842, 970–972.
5. Kármán, Th. von, "Über den Mechanismus des Widerstandes, den ein bewegter Körper in einer Flüssigkeit erfährt," *Göttinger Nachrichten, mathematisch-physikalische Klasse* (1911), 509–517; (1912), 547–556.
6. Hiemenz, K., *Die Grenzschicht an einem in den gleichförmigen Flüssigkeitsstrom eingetauchten geraden Kreiszylinder*, Göttingen Dissertation (1911); *Dingler's Polytechnic Journal*, *326* (1911), 321–324.
7. Rayleigh, Lord, "Aeolian Tones," *Philosophical Magazine*, series 6, *29* (1915), 433–444; also *Scientific Papers* (Cambridge, 1920), VI, 315–325.

8. Gongwer, C. A., "A Study of Vanes Singing in Water," *Journal of Applied Mechanics, 19* (1952), 432–438.

9. Gutsche, F., "Das 'Singen' von Schiffsschrauben," *Zeitschrift des Vereines Deutscher Ingenieure, 81* (1937), 882–883.

10. Kármán, Th. von, "Berechnung der Druckverteilung an Luftschiffkörpern," *Abhandlungen aus dem Aerodynamischen Institut an der Technischen Hochschule Aachen, 6* (1927), 1–17.

11. Reynolds, O., "An Experimental Investigation of the Circumstances which Determine whether the Motion of Water Shall Be Direct or Sinuous and of the Law of Resistance in Parallel Channels," *Philosophical Transactions of the Royal Society of London*, series A, *174* (1883), 935–982; also *Scientific Papers* (Cambridge, 1901), II, 51–105.

12. Newton, I., *Philosophiae Naturalis Principia Mathematica* (London, 1726), Book II.

13. Navier, C. L. M. H., "Mémoire sur les lois du mouvement des fluides," *Mémoires de l'Académie des Sciences, 6* (1823), 389–416.

14. Stokes, G., "On the Theories of the Internal Friction of Fluids in Motion," *Transactions of the Cambridge Philosophical Society, 8* (1845), 287–305.

15. Margoulis, W., "Nouvelle méthode d'essai de modèles en souffleries aérodynamiques," *Comptes rendus de l'Académie des Sciences, Paris, 171* (1920), 997–999.

16. Munk, M. M., "On a New Type of Wind Tunnel," *N.A.C.A. Technical Note* No. 60 (1921); Munk, M. M., and Miller, E. W., "The Variable Density Wind Tunnel of the National Advisory Committee for Aeronautics," *N.A.C.A. Report* No. 227 (1925).

17. Bose, E., and Rauert, D., "Experimentalbeitrag zur Kenntnis der turbulenten Flüssigkeitsreibung," *Physikalische Zeitschrift, 10* (1909), 406–409; Bose, E., and Bose, M., "Über die Turbulenzreibung verschiedener Flüssigkeiten," *ibid., 12* (1911), 126–135.

18. Kármán, Th. von, "Über die Turbulenzreibung verschiedener Flüssigkeiten," *Physikalische Zeitschrift, 12* (1911), 283–284.

19. Hagen, G., "Über den Einfluss der Temperatur auf die Bewegung des Wassers in Röhren," *Abhandlungen der Königlichen Akademie der Wissenschaften zu Berlin, mathematische Klasse* (1854), 17–98; "Über die Bewegung des Wassers in cylindrischen, nahe horizontalen Leitungen," *ibid.* (1869), 1–29.

20. Dryden, H. L., and Kuethe, A. M., "The Measurement of Fluctuations of Air Speed by the Hot-Wire Anemometer," *N.A.C.A. Report* No. 320 (1929); Dryden, H. L., and Kuethe, A. M., "Effect of Turbulence in Wind Tunnel Measurements," *N.A.C.A. Report* No. 342 (1930); Mock, W. C., Jr., and Dryden, H. L., "Improved Apparatus for the Measurement of Fluctuations of Air Speed in Turbulent Flow," *N.A.C.A. Report* No. 448 (1932); Dryden, H. L., Schubauer, G. B., Mock, W. C., Jr., and Skramstad, H. K., "Measurements of Intensity and Scale of Wind Tunnel Turbulence and Their Relation to the Critical Reynolds Number of Spheres," *N.A.C.A. Report* No. 581 (1937).

21. Eiffel, G., "Sur la résistance des sphères dan l'air en mouvement," *Comptes rendus de l'Académie des Sciences, Paris, 155* (1912), 1597–1599.

22. Prandtl, L., "Der Luftwiderstand von Kugeln," *Göttinger Nachrichten, mathematisch-physikalische Klasse* (1914), 177–190.

23. Beaufoy, M., *Nautical and Hydraulic Experiments, with Numerous Scientific Miscellanies* (London, 1834).

24. Froude, W., "Report to the Lords Commissioners of the Admiralty on Experiments for the Determination of the Frictional Resistance of Water on a Surface, under Various Conditions, Performed at Chelston Cross, under the Authority of Their Lordships," *44th Report of the British Association for the Advancement of Science* (1874), 249–255.

25. Prandtl, L., "Grenzschichten in Flüssigkeiten mit kleiner Reibung," *Verhandlungen des dritten internationalen Mathematiker-Kongresses, Heidelberg, 1904* (Leipzig, 1905), 484–491; reprinted by L. Prandtl and A. Betz in *Vier Abhandlungen zur Hydrodynamik und Aerodynamik* (Göttingen, 1927).

26. Kármán, Th. von, "Über laminare und turbulente Reibung," *Zeitschrift für angewandte Mathematik und Mechanik, 1* (1921), 233–252.

27. Pohlhausen, K., "Zur näherungsweisen Integration der Differentialgleichung der laminaren Grenzschicht," *Zeitschrift für angewandte Mathematik und Mechanik, 1* (1921), 252–268.

28. Kármán, Th. von, and Millikan, C. B., "A Theoretical Investigation of the Maximum-Lift Coefficient," *Journal of Applied Mechanics, 2* (1935), 21–27.

29. Taylor, G. I., "Statistical Theory of Turbulence," *Proceedings of the Royal Society of London*, series A, *151* (1935), 421–478; *156* (1936), 307–317.

30. Kármán, Th. de, and Howarth, L., "On the Statistical Theory of Isotropic Turbulence," *Proceedings of the Royal Society of London*, series A, *164* (1938), 192–215.

31. Kolmogoroff, A., "The Local Structure of Turbulence in Incompressible Viscous Fluid for Very Large Reynolds Numbers," *Comptes rendus de l'Académie des Sciences de l'U.R.S.S.*, *30* (1941), 301–305; "Dissipation of Energy in the Locally Isotropic Turbulence," *ibid.*, *32* (1941), 16–18.

32. Heisenberg, W., "Zur statistischen Theorie der Turbulenz," *Zeitschrift für Physik*, *124* (1947), 628–657.

33. Batchelor, G. K., "Kolmogoroff's Theory of Locally Isotropic Turbulence," *Proceedings of the Cambridge Philosophical Society*, *43* (1947), 533–559.

34. Prandtl, L., "Bericht über Untersuchungen zur ausgebildeten Turbulenz," *Zeitschrift für angewandte Mathematik und Mechanik*, *5* (1925), 136–139; "Über die ausgebildete Turbulenz," *Verhandlungen des zweiten internationalen Kongresses für technische Mechanik, Zürich, 1926* (Zurich, 1927), 62–74.

35. Kármán, Th. von, "Mechanische Ähnlichkeit und Turbulenz," *Göttinger Nachrichten, mathematisch-physikalische Klasse* (1930), 58–76; "Mechanische Ähnlichkeit und Turbulenz," *Proceedings of the Third International Congress for Applied Mechanics, Stockholm, 1930* (Stockholm, 1931), I, 85–92.

36. Prandtl, L., "Neuere Ergebnisse der Turbulenzforschung," *Zeitschrift des Vereines Deutscher Ingenieure*, *77* (1933), 105–114.

37. Millikan, C. B., "The Boundary Layer and Skin Friction for Figures of Revolution," *Transactions of the American Society of Mechanical Engineers, Applied Mechanics Section*, *54* (1932), 29–39; Moore, N. B., "The Boundary Layer and Skin Friction for a Figure of Revolution at Large Reynolds Numbers," *The Daniel Guggenheim Airship Institute Publication*, No. 2 (1935), 21–31; also Ph.D. thesis, same title, California Institute of Technology, 1934; Moore, N. B., "Application of Kármán's Logarithmic Law to Prediction of Airship Hull Drag," *Journal of the Aeronautical Sciences*, *2* (1935), 32–34.

38. Sommerfeld, A., "Ein Beitrag zur hydrodynamischen Erklärung der turbulenten Flüssigkeitsbewegungen," *Atti del IV Congresso Internationale dei Matematici, Roma, 1908* (Rome, 1909), III, 116–124.

39. Heisenberg, W., "Über Stabilität und Turbulenz von Flüssigkeitsströmen," *Annalen der Physik*, series 4, *74* (1924), 577–627.

40. Tollmien, W., "Über die Entstehung der Turbulenz," *Göttinger Nachrichten, mathematisch-physikalische Klasse* (1929), 21–44.

41. Lin, C. C., "On the Stability of Two-dimensional Parallel Flows," *Quarterly of Applied Mathematics*, *3* (1945–1946), 117–142, 218–234, 277–301.

42. Dryden, H. L., "Some Recent Contributions to the Study of Transition and Turbulent Boundary Layers," *N.A.C.A. Technical Note* No. 1188 (1947); Schubauer, G. B., and Skramstad, H. K., "Laminar-Boundary-Layer Oscillations and Transition on a Flat Plate," *Journal of Research of the National Bureau of Standards*, *38* (1947), 251–292; also *N.A.C.A. Report* No. 909 (1948).

43. Tani, I., and Hama, F. R., "Some Experiments on the Effect of a Single Roughness Element on Boundary-Layer Transition," *Journal of the Aeronautical Sciences*, *20* (1953), 289–290.

44. Jacobs, E. N., "Preliminary Report on Laminar-Flow Airfoils and New Methods Adopted for Airfoil and Boundary-Layer Investigations," *N.A.C.A. Advance Confidential Report, June 1939;* declassified as *N.A.C.A. Wartime Report* L–345.

45. Lewis, G. W., "Some Modern Methods of Research in the Problem of Flight," *Journal of the Royal Aeronautical Society*, *43* (1939), 771–798.

46. Tani, I., and Mituisi, S., "Contributions to the Design of Aerofoils Suitable for High Speeds," *Report of the Aeronautical Research Institute, Tokyo Imperial University*, No. 198 (1940).

Nikolai E. Joukowski

Osborne Reynolds
(*Courtesy of Manchester University*)

Ernst Mach

W. J. Macquorn Rankine

CHAPTER IV

» Supersonic Aerodynamics

THE subject matter of this chapter is somewhat broader than the title, "Supersonic Aerodynamics," indicates. The chapter deals with the fundamental principles of the aerodynamics of compressible fluids in both supersonic and subsonic flow.

Propagation of Pressure Change: Sound Velocity

Until now we have considered air as a practically incompressible fluid. At moderate speeds the changes of air density and temperature caused by motion are almost negligible. But if we go to higher speeds, the changes of density and temperature caused by compression or expansion of the air become very noticeable. Thus the subject of this chapter is not purely aerodynamics; we may call it *aerothermodynamics*, i.e., a combination of two sciences, fluid mechanics and thermodynamics. The expression aerothermodynamics was first introduced by General G. Arturo Crocco in 1931 (Ref. 1). Later many such words were formed; e.g., in the next chapter, we shall talk about a combination of aerodynamics and elasticity called *aeroelasticity*. We also speak sometimes of *aeroelectronics*, but practical engineers call the corresponding branch of engineering *avionics*.

The essential difference between an incompressible fluid and a compressible fluid is that in the former the propagation of pressure is instantaneous, whereas in the latter the propagation takes place with finite velocity. For example, if we strike the surface of an incompressible fluid, the effect perceived at a great distance is, of course, less than that at a smaller distance, but it reaches even an infinite distance in no time; whereas in a compressible fluid the effect propagates at a finite velocity. The velocity of propagation of a very small pressure change is called the *velocity of sound*. What has the aerodynamics of flight to do with sound? Many laymen ask the question: "Why is it difficult to fly faster than sound?" Really it is not a question of flying faster than sound; it is a question of flying faster than any pressure effect produced in the air can be propagated.

The first man who calculated the propagation of pressure or sound in air was Newton (Ref. 2). His finding was that the square of the speed of propagation is equal to the ratio of the pressure change to the corresponding density change involved in the process. He did not write this result in mathematical form, in spite of the fact that he—perhaps he and G. W. Leibniz—invented calculus; he did not use our present symbols. He did, however, calculate the ratio of pressure to density changes, i.e., in modern language the derivative, $dp/d\rho$, where we suppose that the pressure, p, is a function of the density, ρ. Taking p proportional to ρ, he obtained for the velocity of sound in air a value of 979 feet per second. He compared this result with the velocity of sound measured on an artillery field near London by observation of the time difference between the flash and the sound of a gun fired some distance away. One can assume that the velocity of light is infinite in comparison with that of sound. From the observed time difference Newton concluded that the sound velocity was 1,142 feet per second, which is a correct figure for it at the temperature usually prevailing at sea level.

Newton of course noticed the difference in figures obtained from theory and experiment. Then he followed a method familiar to graduate students, namely, he looked for some excuse to justify the discrepancy. First he remarked that the air was not clean; it always contains some suspended dust particles. He thought that the dust particles would account for a deviation of about 10 percent. Then he thought that moisture content would also act against compression. So he said these two effects together might be responsible for the 17 percent difference. Even very great men sometimes indulge in wishful thinking, which is perhaps a shortcoming of most research men. We must realize, however, that at this time thermodynamics was not known as a science.

Pierre Simon, Marquis de Laplace (1794–1827) (Ref. 3) corrected Newton's computation. The fundamental fact which changed the result is the following: The pressure, p, of a so-called ideal gas is proportional to its density, ρ, in an *isothermal* process, i.e., if the change takes place at constant temperature. On the other hand, if a gas is compressed in a so-called *adiabatic* process, it gets hotter, and if it expands, it gets cooler. We call the process adiabatic if there is no possibility for heat conduction from outside into the gas or vice versa. In this case, we can show that the pressure, p, is proportional to a certain power of the density, ρ^γ, where γ is always larger than one and depends on the number of atoms in the molecule—or, more exactly, on the number of degrees of freedom in which a molecule can store energy. For air, γ is equal to about 1.4, so that the derivative, $dp/d\rho$, is 1.4 times as great as it would be if p were proportional to ρ as Newton assumed. The process involved in sound propagation can be considered, with good approximation, to be adiabatic because the heat conduction is negligible.

Laplace introduced the corresponding correction in Newton's formula for the sound velocity so that the square of the sound velocity became 1.4 times larger than was computed by Newton. This correction explained the discrepancy of about 17 percent between Newton's theory and experiment.

If we consider gas to be made up of particles, i.e., molecules, we find that the velocity of sound is of the same order of magnitude as the velocity of the molecules. As a matter of fact, according to the kinetic theory of gases the mean value of the square of the velocity of molecules is equal to $3p/\rho$. The square of the sound velocity is $\gamma p/\rho$; hence molecular velocity and sound velocity are in the ratio $\sqrt{3/\gamma}$, or 1.46 if $\gamma = 1.4$.

The absolute temperature of a gas is proportional to the kinetic energy of the molecules, and therefore, for a given gas, proportional to the mean value of the square of the velocity of the molecules. Consequently, the square of the sound velocity is also proportional to the absolute temperature of the gas. The sound velocity increases when the temperature increases and decreases when the temperature decreases.

The ratio between the velocity of a body moving through the air and the velocity of sound in the air is called the *Mach number* of the motion. Also the ratio between the velocity of a stream and the sound velocity is called the Mach number of the stream. If the velocity is variable in the field, we call the ratio between the velocity at an arbitrary point and the sound velocity corresponding to the temperature prevailing at that point the *local Mach number*.

Ernst Mach (1838–1916) was a professor of physics in Vienna, who, after teaching physics several years, took over the chair of philosophy. Some people say that his influence in the field of philosophy, especially the theory of knowledge, was perhaps greater than his influence on the progress of physics. At the beginning of this century, his philosophical tenets had considerable effect on scientific thinking.

The concept of the ratio between the velocity of motion and the velocity of sound was used for a long time in scientific literature before the designation Mach number was introduced by Jacob Ackeret in Zurich, just as the term Reynolds number was introduced by Sommerfeld many years after Reynolds' investigations. Ackeret felt the desirability for a special name for this characteristic parameter and chose the name of Mach, who had made pioneer studies of supersonic motion—though not, to be sure, of supersonic flight (Ref. 4).

Mach was also the first man who used the so-called *schlieren method* (method of "striae") for visual observation of supersonic flow. This method is suitable for detecting variations of density, or more exactly, density gradients produced in a gas. It was invented by August Töpler in 1864 (Ref. 5) in order to test the homogeneity of glass in optical instruments.

Fig. 41 shows schematically the use of the schlieren method for the visual investigation of gas flows. We produce a beam of

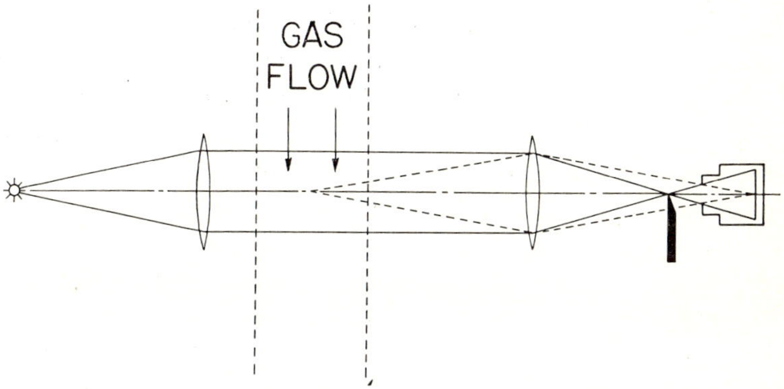

Fig. 41. Optical system for schlieren observation.

parallel light which traverses, perpendicular to the flow direction, the gas flow to be investigated. By means of a lens system we then concentrate the light at a focal point. A camera or a screen for observation is placed behind the focus. Let us now assume that we place a knife edge near the focus. If the knife edge is outside the focus, the field on the screen is bright. If the knife edge covers the focus, the field becomes dark. Let us arrange the knife edge so that it just touches the focus and assume that the density of the air, due to a variable velocity distribution in the gas flow, is nonuniform. In any region where there is a density gradient perpendicular to the direction of the knife edge, the degree of illumination will indicate the gradient, because the density gradient causes a deflection of the light passing through the gas. If the deflection occurs toward the knife edge, the latter will catch a portion of the light; if the deflection is away from the knife edge, the intensity of the light will increase. By changing the direction of the knife edge, one is able to discover the density gradient in any arbitrarily chosen direction. This method is especially suitable for tracing regions in which the density varies rapidly, as, for example, where the air traverses shock fronts.

Propagation of Signals from a Moving Source

Let us now consider the laws of propagation of a pressure impulse produced in a compressible fluid. If the fluid is at rest, the pressure impulse propagates with sound velocity uniformly in all directions, so that the surface which the effect of the impulse reaches at any instant is spherical. If we assume, however, that the source of the impulse is placed in a uniform stream, the impulse will be carried by the stream and at the same time it will propagate relative to the stream with sound velocity. Consequently, the resulting propagation is no longer symmetrical; it is faster in the direction of the stream and slower against the stream. If the stream velocity is equal to the sound velocity, it appears that the effect of the impulse cannot reach every point in the space but is restricted to the half-space bounded by a plane perpendicular to the flow direction. The source of impulse is no longer able to send signals upstream. If the velocity of the stream is supersonic, i.e., superior to the velocity of sound, the effect of the impulse is restricted to a cone whose vertex is the source of the impulse and whose vertex angle decreases from 90° (which corresponds to Mach number equal to 1) to smaller and smaller values as the Mach number of the stream increases. In fact, the trigonometric sine of the half-vertex angle is equal to the reciprocal of the Mach number. The cone which separates the "zone of action" from the "zone of silence" or the "zone of forbidden signals" is called the *Mach cone*, and its half-vertex angle is called the *Mach angle*. Since the trigonometric sine of 30° is equal to one half, the Mach angle 30° corresponds to "Mach 2," i.e., the stream velocity equals twice the sound velocity.

If a source of pressure impulses travels through the air, the conditions are analogous. In Fig. 42(*a*) is shown the source at rest at the point *O*. The concentric circles give the location of the pressure effects due to impulses emitted by the source at equidistant past instants. In Fig. 42(*b*) the source is assumed to move

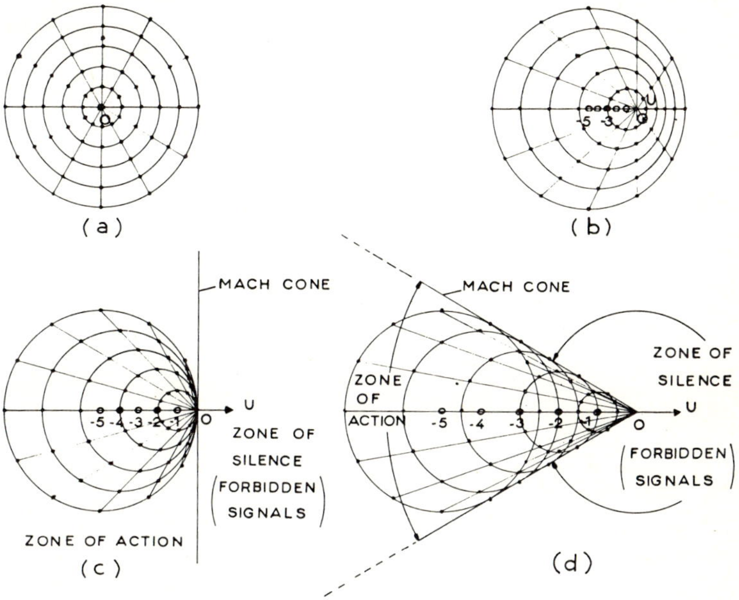

Fig. 42. Point source moving in compressible fluid. (*a*) Stationary source. (*b*) Source moving at half the speed of sound. (*c*) Source moving at the speed of sound. (*d*) Source moving at twice the speed of sound. (From Th. von Kármán, in *Journal of the Aeronautical Sciences*, *14* [1947], 374, by permission of the Institute of the Aeronautical Sciences.)

with subsonic velocity. The small circles indicate the positions of the source at past instants of emission, and the large circles contain the points reached simultaneously by the pressure effects. It is seen that the circles are no longer concentric. Fig. 42(*c*) and (*d*) are corresponding diagrams for sources moving with sonic and supersonic speeds respectively. In the case of a projectile moving with supersonic speed through air at rest we may assume that the main disturbance originates from its vertex. Therefore the effect of the disturbance is restricted to the interior of the Mach cone that moves with the projectile; ahead of the cone the air remains undisturbed. One sees the fundamental difference between the subsonic and the supersonic motion of a body. In subsonic motion, the effect of the disturbance, although decreasing with the distance, reaches every point of the space surrounding the body, whereas in supersonic motion the action is restricted to the inside of the Mach cone. If a projectile goes over your head at a supersonic speed, you hear it only when it is far beyond you. The saying is that nobody ever heard the bullet that killed him—because, before he could perceive the sound, the bullet had already hit him!

Two-dimensional Linearized Wing Theory

Let us now consider the flow pattern produced by a wing moving at a supersonic speed. First we confine ourselves to wings with infinite span, i.e., to the two-dimensional flow problem. If a wing section is thin, we can consider the disturbances caused by the wing to be small. We therefore assume that, in the first approximation, the flow pattern produced by the wing can be built up by superposition of small disturbances emanating from the points of the wing system. The theory of lift and drag for such a wing was first developed by Ackeret (Ref. 6).

Let us consider for simplicity's sake a wing whose section is made up of straight-line segments as shown in Fig. 43. Assume that a uniform and parallel supersonic stream of Mach number M strikes the first element of the wing surface, whose inclination to the stream direction is Θ_1. There are two effects due to the element at L: the flow direction of the stream is changed by the angle Θ_1, and a pressure rise of amount p_1 is produced. The problem is to calculate the magnitude of p_1 if the Mach number and the deflection Θ_1 are known.

We know from the general considerations above that the effects of the impulse p_1 are felt only behind the Mach line LL'. One can show that in two-dimensional flow every fluid particle passing

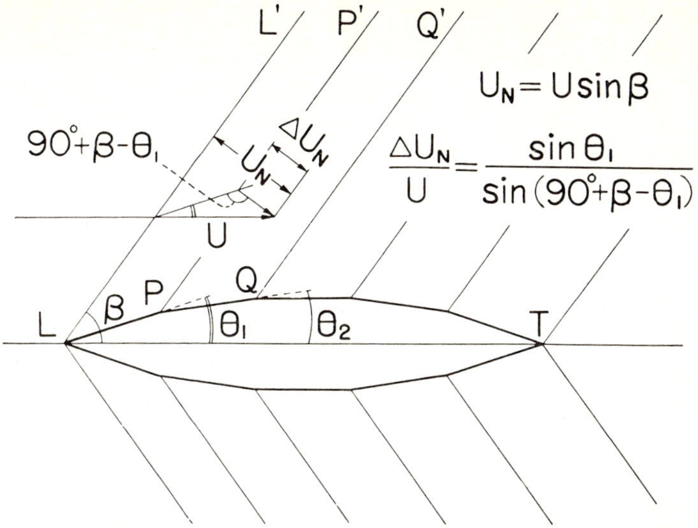

Fig. 43. A symmetrical airfoil section made up of straight-line segments placed with zero angle of attack in a supersonic stream.

through LL' suffers the same deflection, Θ_1, and is subjected to the same pressure rise, p_1. Now we apply the theorem of the equality of pressure force and momentum change. Since the pressure jump occurs perpendicular to the Mach line LL', the pressure rise, p_1, can influence only the velocity component, U_N, whereas the component tangential to LL' must be unchanged. According to the momentum equation, the relation between the pressure rise and the velocity change ΔU_N is $p_1 = \rho U_N \Delta U_N$, where ρ denotes the density of the air.

With the aid of Fig. 43, we can express U_N and ΔU_N in terms of Θ_1 and the angle of inclination β of the Mach line LL'. We remember that $\sin \beta = 1/M$, so that for a thin wing β is always large compared to Θ_1 (except for very large Mach numbers, for which a different theory has to be developed.) Hence, in the formula given in Fig. 43 for the ratio $\Delta U_N/U$, $\sin(90° + \beta - \Theta_1)$ can be replaced by $\cos \beta = \sqrt{M^2 - 1}/M$, and one arrives at the result

$$p_1 = \frac{\rho U^2 \Theta_1}{\sqrt{M^2 - 1}} \cdot$$

In this formula $\sin \Theta_1$ was replaced by Θ_1, which is again correct for small angles.

Let us repeat this calculation at a point farther back along the wing section, assuming the wing to be symmetric and to be placed at zero angle of attack in the stream, as shown in Fig. 43. If the angle of inclination of the succeeding element PQ is Θ_2, the pressure rise caused by this element is $p_2 = \rho U^2 \Theta_2 / \sqrt{M^2 - 1}$. Since Θ_2 is smaller than Θ_1, p_2 is smaller than p_1. We see that the air is accelerated by passing through the Mach line PP', i.e., it expands and experiences a decrease of pressure equal to

$$p_1 - p_2 = \frac{\rho U^2 (\Theta_1 - \Theta_2)}{\sqrt{M^2 - 1}} \cdot$$

In this way the pressure rise relative to ambient pressure decreases as we proceed downstream. It is proportional to the angle of inclination of the surface element and remains positive until we reach the element whose inclination is zero. If we proceed farther, the angle of inclination becomes negative and the pressure falls below the ambient pressure of the stream.

The conclusion is not changed when we increase indefinitely the number of straight-line segments composing the wing surface, i.e., for a wing section with a smooth surface, as shown in Fig. 44. The pressure is constant along the Mach line emanating from a certain point on the surface and has the value $p_0 + \rho U^2 \Theta / \sqrt{M^2 - 1}$, where Θ is the angle of inclination of the tangent at that point to the stream direction and p_0 denotes the ambient pressure. Hence the pressure acting on the front part of the wing is higher and the pressure acting on the rear part is lower than the ambient pressure. The pressure difference between front and rear parts evidently produces a drag. This is a new source of drag, which is additional to the drag components mentioned in Chapter III.

We remember that, at least according to the theory of incompressible nonviscous fluids, the pressures at the front and rear

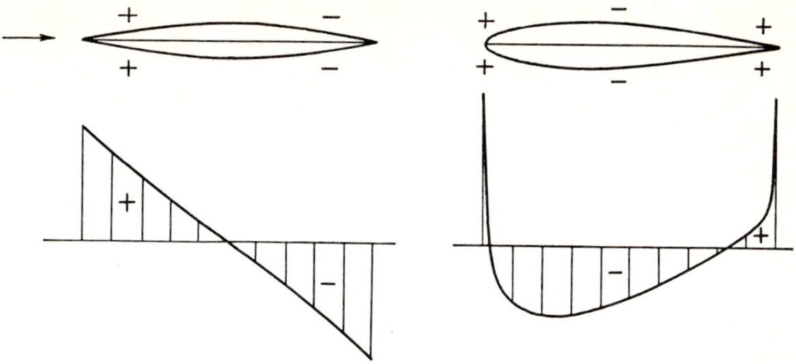

Fig. 44. A symmetrical airfoil section placed with zero angle of attack in a supersonic stream of compressible fluid (*left*) and in a stream of incompressible fluid (*right*). Lower diagrams show the distributions of pressure along the airfoil surface.

parts of a streamlined section balance each other (Fig. 44), as predicted by D'Alembert's theorem. Evidently this theorem does not apply to supersonic flow. For low speeds we usually use a wing section with a blunt nose; the main requirement of streamlining is the sharp trailing edge. For supersonic speeds the blunt nose is rather disadvantageous because of the large angle of inclination that it involves; the sharp trailing edge does not help much because we cannot avoid negative pressure at the rear portion of the section. The essential requirement for supersonic wing sections is a small thickness ratio, i.e., a small value for the ratio between the maximum thickness and the chord length.

We may ask the physical reason for the fact that in a supersonic speed range, even if we neglect skin friction and avoid flow separation, the moving body experiences a drag which has no parallel in subsonic motion. We have seen that, whereas in subsonic motion a pressure change propagates freely in all directions, in supersonic motion the bulk of the action is restricted to the Mach line and in the general three-dimensional case to the surface of the Mach cone. A body produces a system of compression and expansion waves that move with it. This phenomenon reminds the observer of a speedboat when it proceeds with a velocity higher than the velocity of the surface waves and therefore carries with it the waves that it produces. The work that must be done to create and carry these waves is a large part of the total resistance of the boat. With this analogy in mind, we call supersonic drag *wave drag.* The theoretical explanations of both phenomena are based on the same concept. However, when the speedboat goes "on the step," a great part of the wave disappears. Unfortunately, an airplane cannot go "on the step" into the fourth dimension. Some people believe that we only have to get past the sound velocity and everything will be all right. That is, of course, not true.

We now apply the same reasoning to an inclined flat plate, in order to study the laws of the lift produced by a thin wing section. The conclusion is that positive pressure is produced at the lower surface and negative pressure at the upper (Fig. 45). The

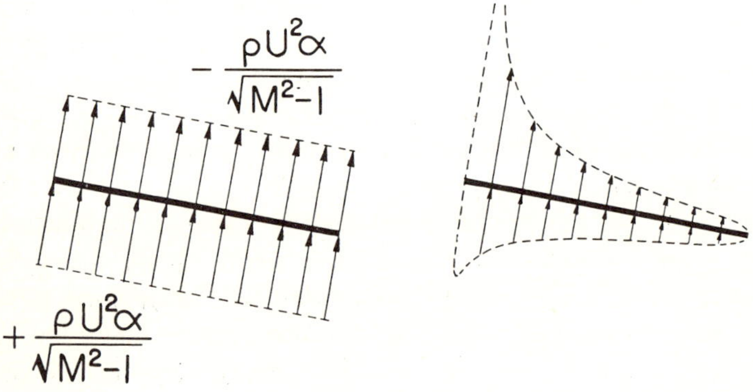

Fig. 45. Pressure distribution around an inclined flat plate in a supersonic stream of compressible fluid (*left*) and in a stream of incompressible fluid (*right*). ρ denotes the density, U the velocity, and M the Mach number of the stream; α denotes the angle of attack.

amounts of the pressure change are $+\rho U^2\alpha/\sqrt{M^2 - 1}$ and $-\rho U^2\alpha/\sqrt{M^2 - 1}$, respectively, where α is the angle of attack. The lift acting on a wing area equal to S is therefore $2\rho U^2\alpha S/\sqrt{M^2 - 1}$ and the lift coefficient C_L, defined as (Lift) $\div \frac{1}{2}\rho U^2 S$, becomes equal to $4\alpha/\sqrt{M^2 - 1}$. According to this formula, for example, C_L is equal to 4α when M is $\sqrt{2}$ or 1.41, and equal to 1.41α when M is 3. The lift coefficient decreases with increasing Mach number. This is also true for the drag coefficient.

If, however, we consider a case in which M is equal to 1, the formula mentioned above gives an infinite value of the lift coefficient (Fig. 46). This is, of course, not correct, and the incorrect

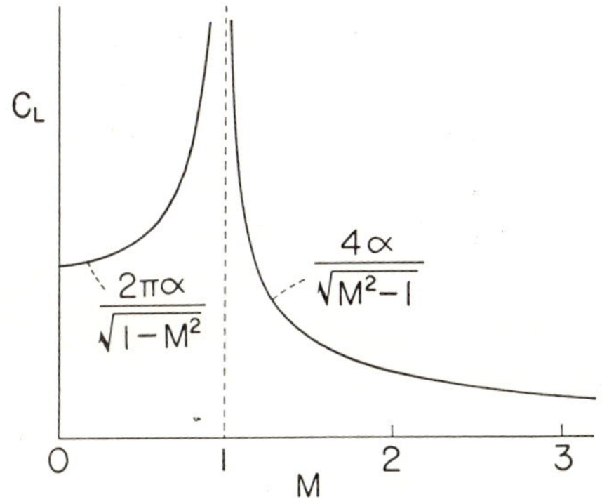

Fig. 46. Lift coefficient of a flat plate at angle of attack α, as a function of the Mach number M, according to the linearized theory.

result is caused by the fact that the simplified theory based on the assumption of infinitely small disturbances, which we call the linearized theory, does not hold for the speed range near the sound speed. As a matter of fact, a linearized wing theory can also be worked out for subsonic flight, in the range of moderately high speeds in which the approximation of incompressible fluids no longer holds, so that the so-called compressibility effects have to be taken into account. For this speed range, we find that the lift coefficient is also a function of Mach number. Prandtl (Ref. 7) and Glauert (Ref. 8) proposed a correction for such speeds. According to their correction formula the lift coefficient for flight at Mach number M is equal to $2\pi\alpha/\sqrt{1 - M^2}$, where $2\pi\alpha$ is the lift coefficient of a flat plate for incompressible fluids ($M \to 0$). It is seen that this theory also fails if M approaches unity, i.e., if we come to nearly sonic flight from below. H. S. Tsien and I (Ref. 9) proposed a somewhat further-reaching correction for compressibility effects on wings, but our method also fails to work when we come near sonic flight or, more precisely, when the local velocity at some point of the wing surface becomes equal or superior to sonic velocity—or, as we say, the local Mach number at some point reaches the value 1. Beyond this limit we have mixed subsonic and supersonic flow regions, and the theory becomes rather involved. Experimentation, for example in wind tunnels, also becomes difficult.

We call the speed range just below and just above the sonic speed—Mach number nearly equal to 1—the *transonic range.* Dryden and I invented the word transonic. We had found that a word was needed to denote the critical speed range of which we were talking. We could not agree whether it should be written with one *s* or two. Dryden was logical and wanted two *s*'s. I thought it wasn't necessary always to be logical in aeronautics, so I wrote it with one *s*. I introduced the term in this form in a report to the Air Force. I am not sure whether the general who read it knew what it meant, but his answer contained the word, so it seemed to be officially accepted.

Before discussing the transonic problem, I would like to say a few words about the linearized theory as applied to three-dimensional flows and also about the effect of finite pressure changes.

Three-dimensional Linearized Theory

As we have seen in Chapter II, wing theory has to deal with two-dimensional problems of wings with infinite span, and with three-dimensional problems of wings with finite span. The same two classes of problems also occur in supersonic wing theory. Ackeret's solution given above is a solution for the two-dimensional problem in linearized form, i.e., under the assumption that the velocities produced by the presence of the wing section are small in comparison with the flight velocity. Further approximations will be mentioned in the next section. In treating three-dimensional problems, most investigators have used linearized theory. With this approximate method, an extensive amount of theoretical information has been gathered, especially in the last ten years, concerning the theory of lift distribution and the calculation of the induced drag and wave drag for various shapes of supersonic wings. This work has been greatly assisted by the fact that the three-dimensional problem of steady supersonic flow can be reduced to that of two-dimensional wave propagation.

The latter problem was very well known before the advent of supersonic flight. Mathematicians and theoretical physicists had done a very good job in this field, so that methods were readily available for the new aerodynamic applications. The analogy with wave propagation in two dimensions is not restricted to wings but also applies to supersonic flow around slender bodies. As a matter of fact, one method well known in the theory of wave propagation, the so-called method of sources, was used in a work of mine, done jointly with Norton B. Moore in 1932, for the calculation of the drag of slender bodies, like projectiles, moving with supersonic speed (Ref. 10). This work appeared before the bulk of papers dealing with three-dimensional supersonic wing theory.

In 1945 a group of American scientists were engaged in collecting German papers and documents produced during the war. The list of German papers was translated into English by an American sergeant. One of my collaborators found, in the list of papers on aerodynamics, one entitled "Resistance of Undernourished Bodies." This was the sergeant's rendering of the German translation of my paper on "Resistance of Slender Bodies."

Among several methods used successfully for the solution of linearized equations of steady supersonic motion I want to mention that of *conical flow*, proposed first by Adolf Busemann in 1942 (Ref. 11). This method tries to build up practically important flow patterns by the superposition of elementary conical flows. The fundamental case of a conical flow is the flow around a circular cone. By solving this relatively simple flow pattern, one finds that in the case of supersonic flow the velocity components are constant along any straight line emanating from the vertex of the cone. In general we call a flow *conical* if it fulfills this condition. By the superposition of such flows, many apparently complicated problems can be solved.

Shock Wave

It was mentioned that the linearized theory of supersonic flow deals only with very small perturbations of a parallel stream and therefore leads to a continuous velocity and pressure field. However, actual flow often behaves differently, and for large pressure changes we need better approximations. If we observe, for example, the supersonic flow past a circular cone, like the "ogive" of a projectile, by optical methods, e.g., by the schlieren method described earlier in this chapter, we see that density changes of considerable magnitude occur abruptly across some surfaces in the flow. We call such a surface a *stationary shock wave*. The origin of this terminology is as follows: We mentioned before that a very small pressure change propagates with sound velocity; however, if we produce a large pressure rise at a point or in a small volume, as in an explosion, the speed of the resulting pressure wave is essentially higher than the velocity of sound, and when the wave passes any point the pressure rises abruptly from the ambient pressure to a rather large value. This phenomenon is called a shock wave, or more exactly a *progressing shock wave*.

The German mathematician G. F. Bernhard Riemann (1826–1866) (Ref. 12) was the first who tried to calculate the relations between the states of gas before and after a shock wave, but he made a mistake, later corrected by W. J. M. Rankine, the British engineer already mentioned in Chapter III (Ref. 13), and the famous French ballistician Pierre Henry Hugoniot (1851–1887) (Ref. 14), independently. Riemann thought that the change would be isentropic, hence that the entropy would remain unchanged through the shock wave. This is not correct. The total energy content (enthalpy) remains unchanged, whereas the entropy always increases through a shock wave. After Rankine and

Hugoniot, shock waves were studied further by a number of scientists. The science of shock waves is very important, not only in aerodynamics, but also in ballistics and in the theory of explosions, detonations, and maybe also in cosmogony. It has really become a separate branch of physical science. If we observe the phenomena while moving with the shock wave, the shock wave appears to be at rest and the air to pass through it. In this case we speak of a stationary shock wave. The velocity of the stream before the shock wave must be supersonic, because the shock wave propagates through air at rest at a velocity greater than sound velocity. Upon transition through the shock wave, velocity, pressure, density, and temperature undergo sudden changes. When the velocity of the oncoming stream is normal to the shock wave, the velocity behind the shock wave becomes subsonic; the direction of the flow is unchanged. If the velocity of the oncoming stream is not normal to the shock wave, the velocity component parallel to the shock wave remains unchanged through the wave front. The velocity component normal to the shock wave changes, however, from supersonic to subsonic magnitude, so that the stream is deflected. I should also note an important theorem discovered by the French mathematician Jacques Hadamard (Ref. 15). According to his theorem, a vortex-free flow ahead of a shock wave can remain vortex-free after passing through the shock wave only when the wave is straight. If the shock wave is curved, it produces vorticity. This is a fact which makes the analysis of motion behind a shock wave rather complicated.

Let us consider again the case of a two-dimensional wing in a supersonic stream. Instead of a Mach line at which the air undergoes an infinitesimally small pressure rise, as in our former linearized theory, we now find, according to more exact theory, a stationary shock wave, i.e., a surface of discontinuity at which —in addition to velocity—density, pressure, and temperature undergo sudden changes. We say the theory is "more exact" because the linearized theory does not reveal such discontinuous changes. However, if we refine the theory still further, taking into account the viscosity and especially the heat conduction of the air, we find that the change may be abrupt but not discontinuous. This is also verified by observation. The shock wave seen in the schlieren picture has, in general, a small but finite thickness, and in very thin air, where the mean free path of the molecules is large, the thickness of the shock wave can be fairly large.

If we observe the behavior of the flow past an airfoil for in-

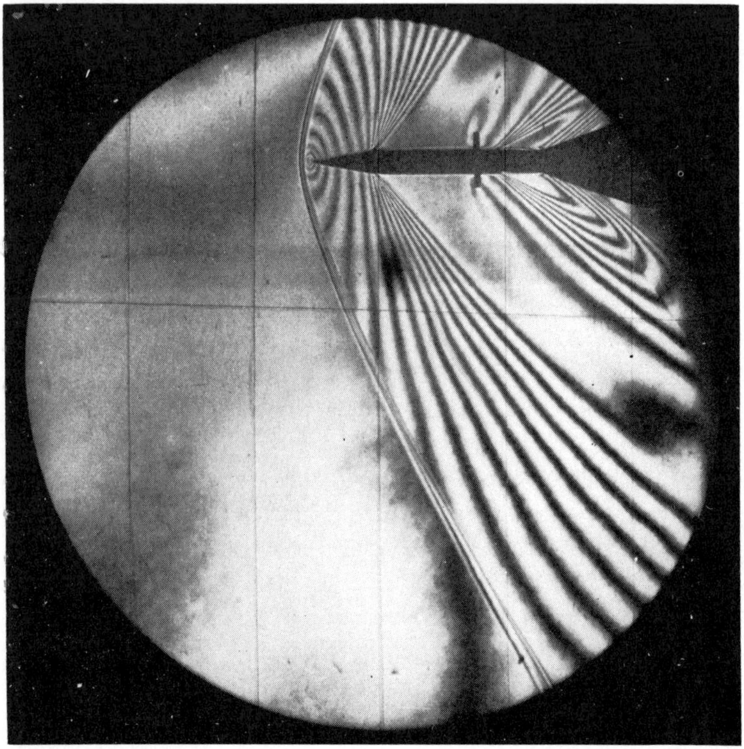

Fig. 47. Detached shock wave of a wedge. Mach number 1.32. The details of the flow field are made visible by the use of an *interferometer*. Light and dark fringes indicate surfaces of equal air density. (Courtesy of Guggenheim Aeronautics Laboratory, California Institute of Technology.)

creasing Mach numbers of the stream, we find that the shock wave attached to the leading edge does not appear immediately after the stream becomes supersonic. First we have a so-called detached shock wave (Fig. 47) at a great distance ahead of the airfoil; the shock wave comes nearer and nearer to the airfoil when the Mach number of the stream increases. At a certain value of the Mach number the shock wave reaches the leading edge and beyond that we find it attached to the leading edge (Fig. 48). (If the leading edge is rounded, the shock wave al-

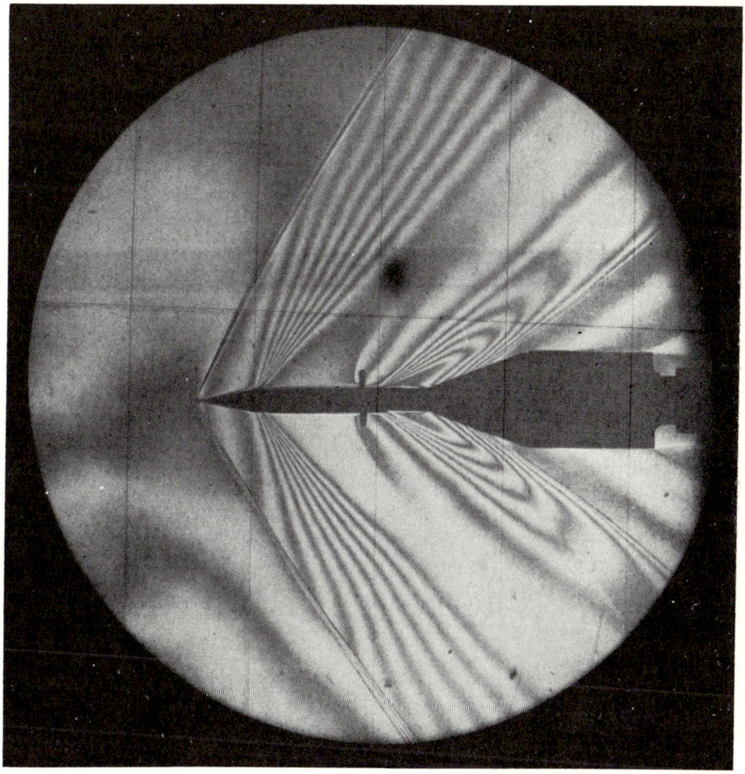

Fig. 48. Attached shock wave of a wedge. Mach number 1.45. The optical technique is the same as in Fig. 47. (Courtesy of Guggenheim Aeronautics Laboratory, California Institute of Technology.)

ways remains detached, but for increasing Mach numbers it is located nearer and nearer to the leading edge.) With a further increase in Mach number the angle of inclination of the attached shock wave decreases and for very large Mach numbers approaches a constant value which is proportional to the half-vertex angle (for air, about 1.2 times the half-vertex angle). Thus, for very high Mach numbers the flow picture is similar to that which Newton imagined in his analysis of air resistance (see Chapter I). According to Newton's assumption, the air proceeds undeflected until it reaches the surface of the body and then is deflected in the direction of the surface. The difference between Newton's flow picture and that we find at very high Mach numbers, called the *hypersonic-speed* range, is that the deflection occurs, not at the surface of the body, but at a surface near to it. This surface appears clearly in Fig. 49. One also finds that in this range the pressure produced at a surface becomes approximately proportional to the square of the angle, as follows from Newton's analy-

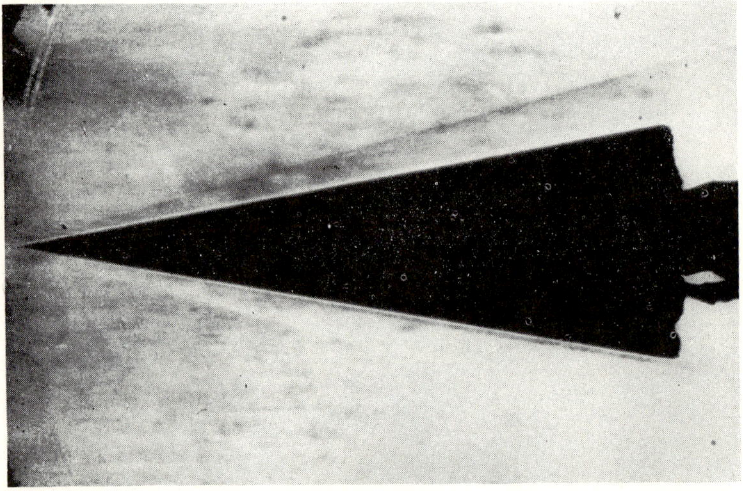

Fig. 49. Schlieren photograph of the attached shock wave of a cone. Mach number 5.9. (Courtesy of Guggenheim Aeronautics Laboratory, California Institute of Technology.)

sis, whereas for moderate Mach numbers the pressure rise is proportional to the angle itself.

We have seen that, according to Ackeret's linearized theory, the deflection of the stream at a concave corner produces a pressure rise, whereas the deflection at a convex corner causes a pressure drop. If we study the same problem by means of the more exact theory, we obtain an abrupt pressure rise through the shock wave emanating from the concave corner (Fig. 50).

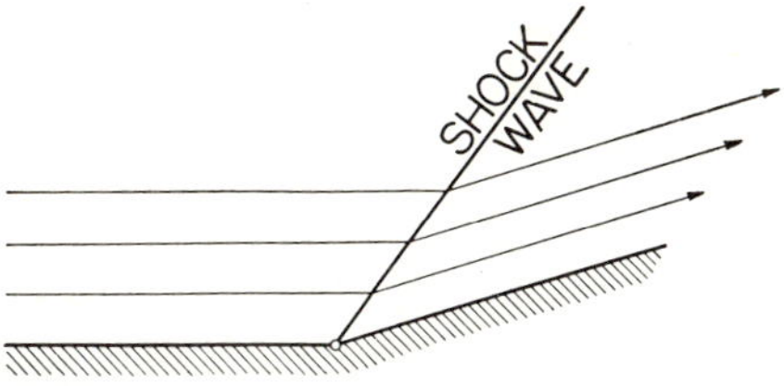

Fig. 50. Flow past a concave corner produces a pressure rise, which is achieved by a shock wave emanating from the corner.

What happens if the flow goes around a convex corner? Both theory and observation show that the air particles go around in a curved path and that the pressure changes gradually from a higher to a lower value (Fig. 51). From the point of view of

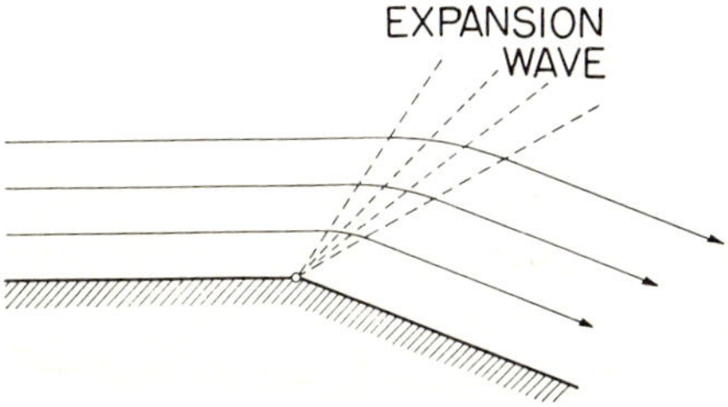

Fig. 51. Flow around a convex corner produces a pressure drop, which is achieved by an expansion wave. The pressure changes gradually from a higher to a lower value through the expansion wave.

fluid mechanics, it is interesting to observe that in supersonic flow the fluid may go around a corner without infinite velocity or separation of flow, whereas it is known that in the subsonic case either the velocity becomes infinite or the flow separates. To be sure, this flow pattern is possible only when the angle of deflection is not too large. The flow pattern of Fig. 52 exhibits both compression shocks and expansion waves.

Fig. 52. Photograph of a conical-headed projectile in flight. Mach number 1.72. A compression shock wave appears at the nose and an expansion wave at the shoulder of the projectile. (Courtesy Ballistics Research Laboratory, Aberdeen Proving Ground, Maryland.)

The fact that there is no "negative shock" in nature, i.e., that if the pressure changes discontinuously the change must involve a positive pressure rise, can be proved by the principles of thermodynamics. A sudden expansion with a sudden temperature drop would mean that the entropy of the gas would decrease without removing heat and without doing external work. This is exactly what is forbidden by the Second Law of Thermodynamics.

I used to illustrate this law to my students by exhibiting two containers in thermic contact, one containing beer, the other tea, both at room temperature. It would certainly be desirable to have the beer cooler and the tea warmer, a process which is perfectly compatible with the law of conservation of energy, i.e., the First Law of Thermodynamics. Unfortunately, the Second

Law makes it wishful thinking, because it would require transfer of heat from a low temperature level to a higher temperature level without the use of mechanical work.

In scientific terms, the impossibility of such a process can be expressed by saying that the entropy would decrease. It can be shown that, in order to make a negative shock possible in a stream, heat would have to be transferred from the lower temperature region behind the expansion wave to the region of higher temperature, against the stream. Thus the expansion shock conflicts with the Second Law of Thermodynamics. The compression shock requires only heat transfer from higher to lower temperature and makes the entropy increase in the gas, as was shown by Rankine and Hugoniot.

Transonic Flight

I want to give here a rather short discussion of the transonic speed range, namely the speed range that extends from just below to just above the sound speed. I want especially to consider the aerodynamics of wings near $M = 1$.

I have already shown in Fig. 46 the lift coefficient of a wing section, according to the linearized theory, in the subsonic and supersonic regions. The lift coefficient becomes infinite if the Mach number approaches 1 from either the subsonic or the supersonic side. This effect does not occur in nature. Instead of increasing to infinity, the lift coefficient reaches a maximum value and then drops, just as in the case of the stall due to increasing angle of attack. As a matter of fact, both phenomena—the decrease of lift coefficient beyond a certain angle of attack and beyond a certain Mach number—are caused by separation of the flow. The question is, What causes the separation in the case of a flow approaching sound velocity?

In order to understand the process, let us consider some flow pictures. Fig. 53 shows the flow pattern of normal subsonic flow around a wing. There is no separation, except a slight tendency to separation near the trailing edge, which may be just a thicken-

Fig. 53. Schlieren picture of the flow pattern of normal subsonic flow. Mach number 0.829. Separation is almost, if not entirely, absent. If present, the separation is slight or manifests itself as a boundary layer of somewhat increased thickness. (Courtesy Guggenheim Aeronautics Laboratory, California Institute of Technology.)

ing of the boundary layer causing a small wake drag. The flow is first accelerated along the upper surface of the wing but does not reach the value of the velocity of sound; after reaching a maximum velocity at a certain point of the surface, the flow is again decelerated.

Fig. 54 shows the flow pattern at a higher subsonic flight speed. The main stream is still subsonic, but near the wing surface there must be supersonic flow because otherwise the shock wave seen there could not appear. Evidently, the flow near the surface (but outside the boundary layer) is accelerated beyond the sound speed. As the flow continues toward the trailing edge, it decelerates, and the transition to subsonic flow occurs by means of a shock. The shock wave is limited in extent at both ends. In the free stream it extends only to a certain distance from the wing surface, since beyond that the flow is no longer supersonic. It also has an end in the boundary layer, since in this layer the velocity decreases to zero at the surface. We notice a slight in-

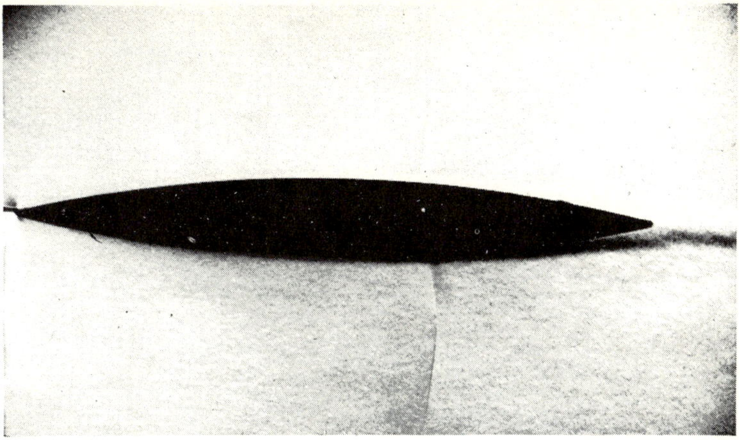

Fig. 54. Schlieren picture of the flow pattern as in Fig. 53 but at a higher speed. Mach number 0.860. A supersonic region is present, and the transition to subsonic flow occurs by means of a shock wave. Definite increase in the thickness of the boundary layer is noticeable, but there is no appreciable separation as yet. (Courtesy Guggenheim Aeronautics Laboratory, California Institute of Technology.)

crease in the thickness of the boundary layer, probably caused by the fact that, owing to the presence of the shock wave, there must be rather rapid increase of pressure along the surface and the boundary layer has to work against pressure rise. We know that exactly this phenomenon, the deceleration of the fluid in the boundary layer by excessive pressure rise, causes flow separation.

In Fig. 55, which refers to a slightly higher Mach number, we see the separation accomplished. By analogy with another case of flow separation we call this effect the *shock stall.* Fig. 55 refers to a case in which the boundary layer is laminar. If the boundary layer is turbulent, it has somewhat more resistance to separation. This mutual effect is known as shock-wave and boundary-layer interaction. The pressure rise produced by a shock wave may cause boundary-layer separation, which in turn reacts on the formation of the shock wave. The problem was first investigated

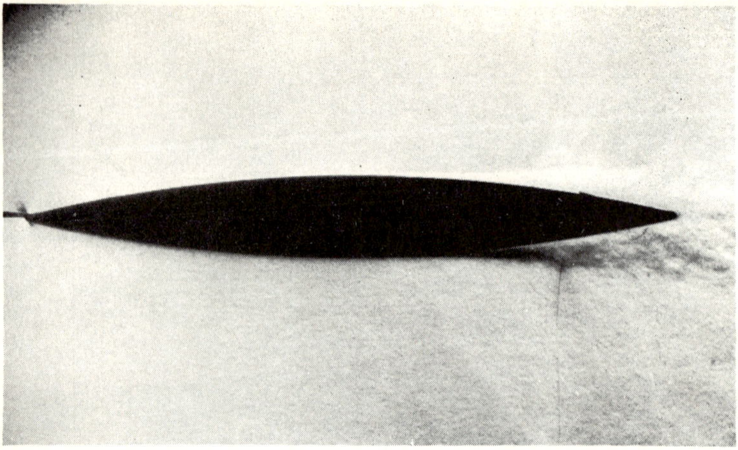

Fig. 55. Schlieren picture of the flow pattern as in Fig. 54 but at a higher speed. Mach number 0.914. Separation of the flow has been accomplished. The boundary layer is laminar in this case. (Courtesy Guggenheim Aeronautics Laboratory, California Institute of Technology.)

by Ackeret, Feldmann, and Rott (Ref. 16) in Zurich and by Liepmann (Ref. 17) at the California Institute of Technology.

The shock stall has two effects on the aerodynamic characteristics of the wing: a decrease of lift and an excessive increase of drag.

Figs. 56 and 57 show schematically the behavior of the lift and drag coefficients of a wing section at a constant angle of attack as functions of the Mach number through the transonic range.

In Chapters II and III we have seen that aerodynamic science

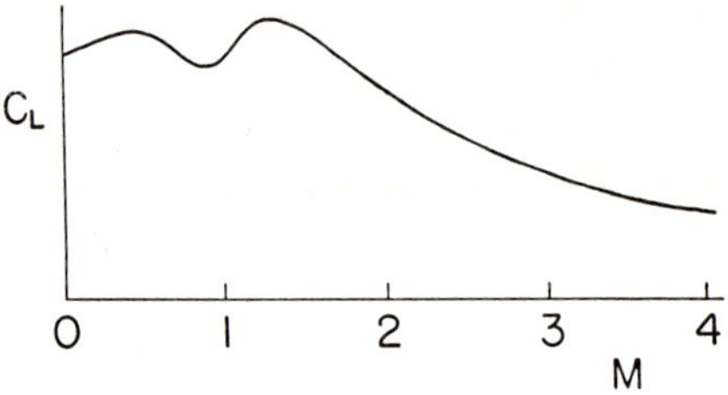

Fig. 56. Lift coefficient C_L of a wing section at a constant angle of attack through the transonic range as a function of the Mach number M.

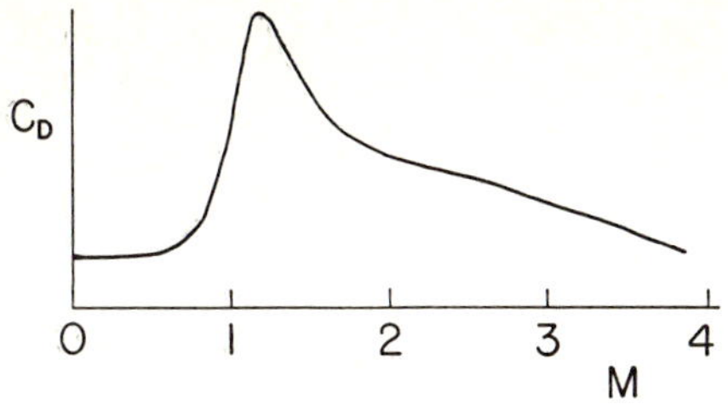

Fig. 57. Drag coefficient C_D of a wing section at a constant angle of attack through the transonic range as a function of the Mach number M.

has succeeded in developing the theory of lift and the theory of drag for incompressible fluids, i.e., for fluid motion at low speeds. These theories enable us to compute, at least to a sufficient approximation, the pressure distribution around the wing section and, by means of the boundary-layer concept, the skin friction acting on the wing surface. In the range of higher subsonic speeds, before we reach the transonic range, the Prandtl-Glauert and Kármán-Tsien theories mentioned above permit reduction of the problem of determining the approximate pressure distribution to that of an incompressible fluid. In the present chapter we have seen that methods are available for computing the lift and drag for supersonic speeds beyond the transonic range. However, the situation is not so favorable in regard to the theory of lift and drag in the transonic range. Solutions of the problem are available only for certain singular cases, certain Mach numbers, and certain wing sections. In general, however, the solution of the flow equations requires extremely cumbersome calculations with no certainty that the results are exact.

In this situation, a similarity consideration, which I proposed and called the *transonic similarity rule*, does good service, since it allows the transfer of experimental results from one case to another (Ref. 18). Suppose we have two thin wing sections that are geometrically similar in the sense that they would become identical if the scale of thickness were changed. For example, we might compare two wing sections, one with 3 percent and the other with 6 percent maximum thickness, the distribution of the ordinates expressed in terms of the maximum ordinate being the same. We find from the consideration of the flow equations that, so far as two-dimensional flow is concerned, the flow pattern must be similar if the ratio $t^{\frac{1}{3}}/\sqrt{|1 - M^2|}$ has the same value, where t is the maximum thickness ratio and M is the Mach number. Hence, if we have a measurement of pressure distribution, lift coefficient, or drag coefficient for one of the wing sections as functions of the Mach number, we are able to compute the corresponding quantities for the other similar wing section with a different thickness ratio. The predictions from this similarity rule agree very well with experiments. It is also found that the similarity rule is approximately correct even when relatively weak shock waves appear in the flow.

It is interesting to know that both the Prandtl-Glauert theory for subsonic speeds and the Ackeret theory for supersonic speeds yield analogous similarity rules for their respective speed ranges. In two-dimensional flow, the corresponding rule would state that the flows are similar when the ratio $t/\sqrt{1 - M^2}$ or $t/\sqrt{M^2 - 1}$ remains constant. The first ratio is a real number for values of M less than unity and the second for M greater than unity.

The appearance of shock waves and the phenomenon of shock stall cause significant changes in the behavior of an airplane flying through the transonic speed range, which, with some simplifications, can be summarized as follows:[1]

a) Unexpected changes can occur in the trim of the airplane. Suppose, for example, that the wing suffers shock stall before the tail does. (This is very possible, since both the thickness ratio and the angle of attack of the wing may be greater than the corresponding parameters of the tail surface.) Evidently, the sudden decrease of the lift at the wing will cause a strong nose-heavy moment. Or, because of the appearance of a shock wave at the upper surface of the wing, the point of action of the re-

[1] The author is indebted to W. Lavern Howland of the Lockheed Aircraft Corporation for this concise formulation of the manifold problems of transonic flight.

sulting lift may be suddenly displaced, disturbing the relative location of lift and gravity forces.

b) Various severe disturbances can occur in the maneuverability of the aircraft. Sometimes the pilot finds that his elevator or rudder is utterly ineffective; he moves the stick or the rudder pedals, but the airplane fails to respond. This can be explained by the shock stalling of the fixed horizontal or vertical surfaces, in the presence of which the control surface moves in the wake and has no effect. At another time the pilot may find that the control surface is "frozen"; apparently the aerodynamic hinge moment has become so large that he is unable to overpower it. No complete explanation is known for this phenomenon; perhaps it has to do with the location of the shock wave. Finally some pilots say that they observe a shift of the control surfaces at a certain Mach number on a given airplane: the rudder, elevator, or aileron may suddenly leave its neutral position and jump to a deflected position without any action of the pilot.

c) Vibrations of the tail, or even of the whole aircraft, are often observed. Presumably in the mixed subsonic-supersonic flow over the wing the positions of the shock waves are not well defined; they may move back and forth. It has also been observed that, when shock waves are produced on both the upper and lower surfaces of the wing or tail, they may move in opposite phase, which apparently makes the wake oscillate, and this oscillation is transferred to the wing or tail.

When such difficulties first appeared in flight, they were described as "compressibility troubles."

I remember a conference in 1941 when the Lockheed Aircraft Corporation built one of the first airplanes that reached Mach numbers greater than 0.7. The airplane became nose-heavy during a dive, and the oscillation originating in the tail unit shook the whole airplane with great violence. A number of "aerodynamic doctors" were called for consultation and diagnosis of the disease. Some said it was ordinary wing flutter, a kind of oscillation that we will discuss in Chapter V. I was one of the "doctors," and I voted for shock stall—and I think I was correct. Indeed, subsequent investigation by the Lockheed Corporation showed that the maximum lift coefficient that could be reached without tail oscillation decreased with increasing Mach number. This may have been the first case of transonic difficulties in actual flight. Spitfires had the problem in operations over England ~1940.

I well remember this period when designers were rather frantic because of the unexpected difficulties of transonic flight. They thought the troubles indicated a failure in aerodynamic theory. I thought we had to expect compressibility effects, since the air has always been compressible. It is rather remarkable that we could go as far as we did with a theory based on the assumption that the air can be treated as an incompressible fluid.

From the practical point of view of minimizing transonic troubles, an increase in the size of the control surfaces or improvement in their efficacy by special devices might be recommended. Also an increase in the force available to the pilot for the operation of the control surfaces, by means of so-called booster controls, is often necessary. Furthermore, excess propulsive power is desirable to permit quick passage through critical speed ranges; it has been noticed, indeed, that some of the dangerous effects are reduced to a slight jerk or lurch if the airplane passes quickly through the transonic range.

Sweptback Wings

There is an effective method to postpone to higher Mach numbers the troubles connected with transonic flight. Everyone is familiar with pictures showing airplanes having wings with *sweepback*, i.e., wings whose leading edges form a considerable angle relative to the perpendicular to the flight direction. The basic theoretical idea underlying the use of such wing planforms can be described as follows: Assume that a wing with constant section and infinite span moves through the air in a direction oblique to its span. We may say that the motion of the wing is composed of a motion normal to the span and a sideslipping motion along the span. If we neglect frictional forces, the latter part of the motion should have no influence on the forces acting on the wing. So we conclude that the flow pattern relative to the wing is determined by an "effective Mach number," which corresponds to the component of the flight speed perpendicular to the span. If, for example, the sweepback angle is 45°, the effective Mach number is about 70 percent of the flight Mach number, so that the critical value of the latter, where transonic troubles appear, would be raised about 40 percent.

Of course, things are not really so simple. First, in the case of sweptback wings of finite span, the theory does not apply to the center part or to the wing tips; secondly, friction and boundary layer have disturbing effects. Nevertheless, the rise of drag and the change in trim usually connected with a Mach number approaching unity are postponed to higher Mach numbers. The benefit in the increase of Mach number is about half of what one would expect according to the simple theory sketched above.

The aerodynamic properties of sweptback wings were treated by Busemann at the Volta Congress for High Speed Flight held in Rome in 1935 (Ref. 19). I remember that, at the banquet of the congress, General Crocco, the organizer of the congress and a man of far-reaching vision, improvised a drawing of an airplane on the back of the menu card. He called it, jokingly, Busemann's airplane; it had sweptback wings and tail, and even its propeller blades were sweptback. Busemann, however, considered the behavior of sweptback wings only in supersonic flight and based his computation of lift and drag on the linearized theory. It is said that Albert Betz first suggested that sweepback might be useful in postponing the transonic effects to higher flight Mach numbers. The suggestion was followed up by wind-tunnel research men and airplane designers. In this country the theory of sweepback was independently discovered in 1945 by Robert T. Jones (Ref. 20).

When I went to Germany with a group of scientists and engineers in 1945, we found, in the deserted Volkenrode Laboratory near Braunschweig, wind-tunnel models of an airplane with sweptback wings and pertinent high-Mach-number wind-tunnel data. George Schairer, eminent chief of the technical staff of the Boeing Aircraft Company, was a member of my group. He had heard of Robert Jones's ideas about sweepback, but the Volkenrode data were the first experimental results he had seen. It is related that Schairer wired back to his home office: "Stop the bomber design" and that this led to the birth of the present B-47 airplane, the first bomber with sweptback wings in this country.

An interesting version of a sweptback wing is the so-called crescent wing, in which the sweepback angle varies along the span of the wing. The sweepback angle is large in the center part where the wing thickness is large and smaller at the outer part of the wing where the wing is thinner.

The delta wing takes advantage of both a large sweepback angle and a small thickness ratio. A small thickness ratio at the center part is preserved by the use of large chord lengths. Since at high speed, transonic and supersonic, the unavoidable profile drag is relatively large in comparison to induced drag, a small aspect ratio is acceptable. The large chord allows relatively large volume inside the wing, which can be used for the storage of fuel or for other loads. Furthermore, one important feature of the delta planform is that the displacement of the center of pressure by transition from subsonic to supersonic flight is smaller than for conventional planforms. Most delta-wing airplanes have only vertical stabilizers. The delta wing can be made longitudinally stable without a horizontal stabilizer, and elevators and ailerons can be arranged at the trailing edge of the wing.

Piercing the Sonic Barrier

At present the problem of "piercing the sonic barrier" appears to be essentially one of propulsion. If enough propulsive force is available to overcome the drag increase occurring at and immediately before the sonic barrier, so that the airplane can pass quickly through the critical speed range, no specific difficulties need be expected. Probably it should be easier to fly an airplane in the supersonic speed range than in the transition range between subsonic and supersonic speed.

Thus the situation is somewhat analogous to that which prevailed in the beginning of this century, when the Wright brothers were able to prove the possibility of powered flight because they had a light engine with sufficient propulsive power. If we have suitable engines, supersonic flight will be rather common. Until just recently the piercing of the sonic barrier in level flight has been accomplished only by the use of rather uneconomical propulsive devices, such as rockets and ramjets, with a very high fuel consumption. Experimental airplanes like the X–1 and the Skyrocket have rockets which are good for only a few minutes' flight or use turbojets with afterburners, but at the moment of writing there are few airplanes that can fly at supersonic speeds for half an hour. If you read in the newspaper that an airplane "went through the sonic barrier," it often means that it did so by diving, in which case gravity supplemented the inadequate propulsive force.

There is a strange phenomenon connected with these diving stunts which I want to mention. Assume that the airplane approaches an observer at subsonic speed, makes a dive reaching supersonic speed, then recovers from the dive and continues in flight at subsonic speed again. In such a case the observer on the ground frequently hears two loud booming sounds, rather closely following one another: "Boom, boom!" Some scientists have offered explanations of the origin of the double boom. Ackeret in Zurich (Ref. 21) and Maurice Roy in Paris (Ref. 22) have both suggested that the booms are due to the piling up of sound impulses—such as engine noise—emitted during the periods in which the airplane passed through the sonic speed. If the airplane is moving toward the observer, the noise emitted from the airplane will reach the observer in a shorter interval than the interval in which it was emitted. Thus some piling up of the sound impulses always takes place, provided the sound source moves toward the observer. However, if the sound source moves at a speed near sound velocity, the piling up becomes infinitely intensified. This becomes evident if one considers that all sound emitted from a source moving exactly with sound velocity straight toward the observer would reach the latter in one short moment, namely, when the sound source arrived at the location of the observer. The reason is that sound and sound source would be travelling at the same speed. If the source moved for a period of time with supersonic speed, the sequence of received and emitted sound impulses would be reversed; the observer would perceive signals emitted at a later time before he perceived those emitted earlier.

The process of the double boom, according to this theory, can be illustrated by the diagram shown in Fig. 58. We assume that

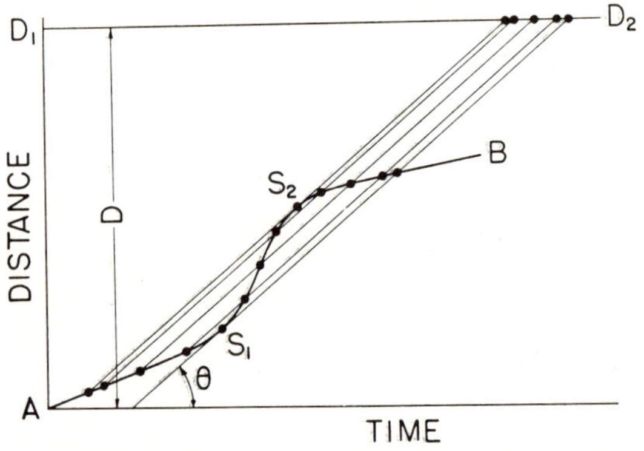

Fig. 58. Distance-time diagram of an airplane flying at variable speed. The parallel lines with the angle of inclination θ indicate the propagation of sound.

the airplane moves straight toward the observer but at variable speed. The curve *AB* shows the displacement of the airplane as a function of the time. The angle of inclination of the tangent to the curve indicates the instantaneous speed of the airplane. The parallel lines shown in the diagram indicate the propagation of sound; the angle of inclination, θ, of these lines corresponds to sound velocity. The velocity of the airplane is first subsonic in the region A to S_1, then supersonic in the region S_1 to S_2, and finally subsonic again in the region S_2 to B. If the observer is at an initial distance D, the points indicated on the horizontal line D_1D_2 correspond to the sequence of sound impulses as perceived by him. We see that the noise emitted by the airplane during its second passage through the sonic barrier (point S_2) reaches the observer earlier than the sound emitted during its first passage (point S_1). At these two instants, the observer perceives, in an infinitely small interval, the impulses emitted during a finite period. Hence he hears an explosionlike boom. Between the two booms he perceives simultaneously three impulses emitted at different times by the airplane.

Fig. 59 shows schematically the intensity of noise we may expect in this simplified case. It should be mentioned that the

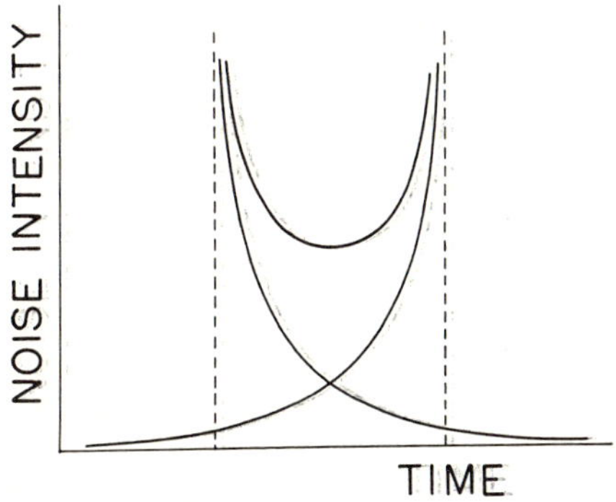

Fig. 59. Schematic representation of noise intensity received by the observer.

piling up of sound impulses in the case of an approaching sound source is the same as the process known as the *Doppler effect;* however, the description of the latter effect is usually restricted to the change in pitch of the tone connected with the piling-up process. It is difficult to calculate the intensity of the perceived noise, since this depends on the mechanism of sound formation, which is not well known. Also the process is complicated by the shape of the trajectory, by possible echoes, and also by shock waves which occur on various parts of the airplane during the flight and whose energy is transformed into sound waves after the airplane decelerates. Some recent papers on the subject attribute the entire phenomenon of the double—sometimes triple—boom observed in supersonic dives to these shock waves.

The problem of "piercing the sonic barrier," or the "sonic wall," seems to appeal to the imagination of the general public (an English motion picture entitled "Breaking the Sound Barrier" gives some idea of the problems connected with flight through Mach 1); pilots and engineers discuss the problem both seriously and jokingly. The following "scientific report" of a transonic flight shows a nice combination of technical knowledge and poetic license (Ref. 23):

We were slipping smoothly through the air at 540 mph. I'd always liked the little XP–AZ5601–NG because of her simple controls and that Prandtl-Reynolds meter tucked away in the upper right corner of the panel. I checked over the gages. Water, fuel, rpm., Carnot efficiency, groundspeed, enthalpy. All OK. Course 270°. Combustion efficiency normal at 23 percent. The good old turbojet was rumbling along as smoothly as always and Tony's teeth were barely clattering from the 17 buckets she'd thrown over Schenectady. Only a small stream of oil was leaking from the engine. This was the life.

I knew the engine in my ship was good for more speed than we'd ever tried. The weather was so fair, the sky so blue, the air so smooth, I couldn't resist letting her out a little. I inched the throttle forward a notch. The regulator only hunted a trifle and everything was steady after five minutes or so. 590 mph. I pushed the throttle again. Only two nozzles clogged up. I pushed the small-slot cleaner. Open again. 640 mph. Smooth. The tailpipe was hardly buckled at all—there were still several square inches open on one side. My fingers were itching on the throttle and I pushed it again. She worked up to 690 mph., passing through the shaft critical without breaking a single window in the ship. It was getting warm in the cockpit so I gave the vortex refrigerator a little more air. Mach 0.9! I'd never been that fast before. I could see a little shocklet outside the port window so I adjusted the wing shape and it disappeared.

Tony was dozing now and I missed the smoke from his pipe. I couldn't resist letting the ship out another notch. In ten minutes flat we leveled off at Mach 0.95. Back in the combustion chambers the total pressure was falling like hell. This was the life! The Kármân indicator showed red but I didn't care. Tony's candle was still burning. I knew gamma was down but I didn't give a damn.

I was dizzy with the thrill. Just a little more! I put my hand on the throttle but just at that moment Tony stretched and his knee struck my arm. The throttle jumped up a full ten degrees! Crash! The little ship shuddered from stem to stern and Tony and I were thrown into the panel by the terrific deceleration. We seemed to have struck a solid brick wall! I could see the nose of the ship was crushed. I looked at the Mach meter and froze. 1.00! My God, I thought in a flash, we're on the peak! If I don't get her slowed down before she slips over, we'll be caught in the decreasing drag! I was too late. Mach 1.01! 1.02! 1.03! 1.04! 1.06! 1.09! 1.13! 1.18! I was desperate but Tony knew what to do. In a flash he threw the engine into reverse! Hot air rushed into the tailpipe, was compressed in the turbine, debusted in the chambers, expanded out the compressor. Kerosene began flowing into the tanks. The entropy meter swung full negative. Mach 1.20! 1.19! 1.18! 1.17! We were saved: She crept back, she inched back, as Tony and I prayed the flow divider wouldn't stick. 1.10! 1.08! 1.05! Crash! We had struck the other side of the wall! Trapped! Not enough negative thrust to break back through! As we cringed against the wall, the tail of the little ship crushed, Tony shouted, "Fire the JATO units!" But they were turned the wrong way! Tony thrust his arm out and swung them forward, the Mach lines streaming from his fingers. I fired them! The shock was stunning. We blacked out.

I came to as our gallant little ship, ragged from stem to stern, was just passing through Mach zero. I pulled Tony out and we slumped to the ground. The ship decelerated off to the east. A few seconds later we heard the crash as she hit the other wall.

They never found a single screw. Tony took up basket weaving and I went to M. I. T.[2]

[2] Reproduced by permission of *Aviation Week* and the author, Prof. C. D. Fulton.

References

1. Crocco, G. A., "Sui corpi aerotermodinamici portanti," *Rendiconti della Accademia Nazionale dei Lincei*, series 6, *14* (1931), 161–166; "Flying in the Stratosphere," *Aircraft Engineering*, *4* (1932), 171–175, 204–209.

2. Newton, I., *Philosophiae Naturalis Principia Mathematica* (London, 1726), Book II, Proposition 50.

3. Laplace, P. S. de, "Sur la vitesse du son dans l'air et dans l'eau," *Annales de chimie et de physique*, series 2, *3* (1816), 238–241.

4. Mach, E., and Salcher, P., "Photographische Fixierung der durch Projectile in der Luft eingeleiteten Vorgänge," *Sitzungsberichte der Wiener Akademie der Wissenschaften*, Abt. II, *95* (1887), 764-780; Mach, E., and Mach, L., "Weitere ballistisch-photographische Versuche," *ibid.*, Abt. IIa, *98* (1889), 1310–1326; Mach, E., and Salcher, P., "Optische Untersuchung der Luftstrahlen," *ibid.*, Abt. IIa, *98* (1889), 1303–1309.

5. Töpler, A., *Beobachtungen nach einer neuen optischen Methode* (Bonn, 1864).

6. Ackeret, J., "Luftkräfte auf Flügel, die mit der grösserer als Schallgeschwindigkeit bewegt werden," *Zeitschrift für Flugtechnik und Motorluftschiffahrt*, *16* (1925), 72–74.

7. Prandtl, L., Über Strömungen, deren Geschwindigkeiten mit der Schallgeschwindigkeit vergleichbar sind," *Journal of the Aeronautical Research Institute, Tokyo Imperial University*, No. 65 (1930).

8. Glauert, H., "The Effect of Compressibility on the Lift of an Aerofoil," *Proceedings of the Royal Society of London*, series A, *118* (1928), 113–119.

9. Tsien, H. S., "Two-dimensional Subsonic Flow of Compressible Fluids," *Journal of the Aeronautical Sciences 6* (1939), 399–407; Kármán, Th. von, "Compressibility Effects in Aerodynamics," *ibid.* *8* (1941), 337–356.

10. Kármán, Th. von, and Moore, N. B., "Resistance of Slender Bodies Moving with Supersonic Velocities, with Special Reference to Projectiles," *Transactions of the American Society of Mechanical Engineers*, *54* (1932), 303–310 (APM–54–27).

11. Busemann, A., "Infinitesimale kegelige Überschallströmung," *Jahrbuch der deutschen Akademie der Luftfahrtforschung* (1942), 455–470.

12. Riemann, B., "Über die Fortpflanzung ebener Luftwellen von endlicher Schwingungsweite," *Abhandlungen der Königlichen Gesellschaft der Wissenschaften zu Göttingen, mathematisch-physikalische Klasse, 8* (1858–59), 43–65; also *Gesammelte Mathematische Werke* (Leipzig, 1876), 145–164.

13. Rankine, W. J. M., "On the Thermodynamic Theory of Waves of Finite Longitudinal Disturbance," *Philosophical Transactions of the Royal Society of London*, series A, *160* (1870), 277–286; also *Miscellaneous Scientific Papers* (London, 1881), 530–543.

14. Hugoniot, H., "Mémoire sur la propagation du mouvement dans les corps et spécialement dans les gases parfaits," *Journal de l'École Polytechnique, Paris, 57* (1887), 1–97; *59* (1889), 1–125.

15. Hadamard, J., *Leçons sur la propagation des ondes et les équations de l'hydrodynamique* (Paris, 1903).

16. Ackeret, J., Feldmann, F., and Rott, N., "Untersuchungen an Verdichtungsstössen und Grenzschichten in schnell bewegten Gasen," *Mitteilungen aus dem Institut für Aerodynamik an der Eidgenössischen Technischen Hochschule in Zürich*, No. 10 (1946).

17. Liepmann, H. W., "The Interaction between Boundary Layer and Shock Waves in Transonic Flow," *Journal of the Aeronautical Sciences, 13* (1946), 623-637.

18. Kármán, Th. von, "The Similarity Law of Transonic Flow," *Journal of Mathematics and Physics, 16* (1947), 182–190.

19. Busemann, A., "Aerodynamischer Auftrieb bei Überschallgeschwindigkeit," *Convegno di Scienze Fisiche, Matematiche e Naturali; Tema: Le Alte Velocità in aviazione, Roma, 1935* (Rome, 1936), 315–347.

20. Jones, R. T., "Wing Plan Forms for High-Speed Flight," *N.A.C.A. Report* No. 863 (1947).

21. Ackeret, J., "Akustische Phänomene bei hohen Fluggeschwindigkeiten," *Neue Zürcher Zeitung*, Sept. 18, 1952.

22. Roy, M., "A propos du gong sonique," *Comptes rendus de l'Académie des Sciences, Paris, 235* (1952), 756–759.

23. Fulton, C. D., "Through the Sonic Wall," *Aviation Week, 51* (Dec. 12, 1949), 56.

CHAPTER V

» *Stability and Aeroelasticity*

THE critical problems in achieving human flight were to develop light engines for propulsion and efficient wing surfaces for sustentation and to secure steady flight or *stability* of the airplane. In this chapter we are concerned with the last-mentioned problem. Before we enter on the subject, it seems desirable to make clear what we mean by the word stability. In relation to a flying machine we are interested in the stability of a motion, but for better understanding of the concept we begin with a discussion of the stability of equilibrium.

Static Stability

Consider the simple case of a solid body suspended at a point above its center of gravity like a pendulum. If we deflect the body by a small angle, the moment of the gravity force tends to restore it to its original position. We say that the equilibrium is stable. On the other hand, it is clear that the pendulum is still in equilibrium if it is inverted, so that its center of gravity is directly above the point of suspension. In this case, however, the equilibrium is unstable, because the moment resulting from a small deflection tends to increase the deviation from the original position.

For a ship floating on water, the stability of equilibrium is an important condition which must be fulfilled. We know that a ship is in equilibrium if the resultant of the lift due to buoyancy passes through the center of gravity. If we deflect the ship as shown in Fig. 60, the lift acts through the center of gravity, *B*, of the displaced water while the gravity force acts through the center of gravity of the ship, *G*. These two forces create a moment which tends to restore the ship to its original position of equilibrium, provided that the center of gravity, *G*, is located below the point *M*, where the vertical line through point *B* intersects the centerline of the ship. The point of intersection, *M*, is called the *metacenter*. If the center of gravity is above the metacenter, the equilibrium is unstable and the moment resulting from the deflection tends to increase the angle of inclination until the ship is capsized.

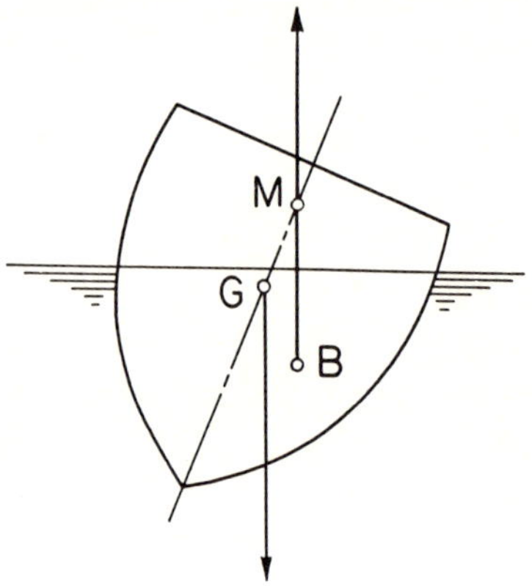

Fig. 60. Cross section of a deflected ship.

This sort of consideration was applied to the stability of flight by the early investigators. In those times only flight based on aerostatics, i.e., balloon flight, had been actually accomplished; thus the early investigators did not recognize the difference between the stability of equilibrium and the stability of motion. We find, for example, that one flight enthusiast in a semiscientific article suggested that the stability of the flying bird depends on the form of its belly, whether its center of gravity is located below the geometrical metacenter of the body. Other investigators, however, based their studies of stability on sounder principles: the airplane in steady flight was considered as a system acted upon by gravity and lift forces. The gravity force acts at the center of gravity of the airplane, while the lift force set up by a flat wing surface acts at about a quarter of the chord back from the leading edge. An obvious condition for equilibrium in steady flight is that the moments of the lift forces acting on the wing and the tail, taken about the center of gravity of the airplane, must balance, the larger force due to the wing being balanced by the smaller force due to the tail, which has a larger moment arm. This is the condition of trim. In order for the equilibrium to be stable, however, a second condition is required, namely, that if the equilibrium is disturbed the moment resulting from the lift acting on the wing and tail must be such that it tends to restore the airplane to the original position. If this condition is fulfilled, we say that the airplane is *statically* stable. Pénaud first (1871) recognized the importance of the tail in securing static stability (Ref. 1). In particular he found that a stabilizing moment can be produced if the wing and tail form a so-called *longitudinal dihedral*, so that the tail is set at an angle of attack smaller than that of the wing. He demonstrated his conclusion with a small model fitted with a propeller driven by rubber bands (Fig. 12, p. 23).

If the wing surface is not flat but curved, the problem becomes somewhat more complicated, because the lift, as mentioned before, has two components: one produced by the curvature, the other produced by the angle of attack. If the wing has the shape of a circular arc, the first component acts at the midpoint of the chord, whereas the second component acts at the front quarter-chord point. Consequently, the point of action of the total lift acting on the wing itself moves as the angle of attack changes; at zero angle of attack it is at the center of the chord; it moves forward as the angle of attack increases. This effect was known to early investigators. They liked to express the stability condition in a form familiar to shipbuilders by generalizing the concept of the metacenter. — NEUTRAL POINT

To maintain static stability in the airplane, it is not absolutely necessary to have a tail. The idea of a tailless airplane is attractive since the tail represents additional weight and drag. The first design of a tailless airplane dates back to 1910 when one was proposed and built in England by J. W. Dunne. More recently the noted American airplane designer John K. Northrop became interested in developing large airplanes of tailless design, which he called "flying wings." W. R. Sears and I helped him in the study of the aerodynamic requirements for stability without the use of a tail. It appears that the wing has to be given a considerable amount of sweepback and twisted so that the angle of attack at the tips is smaller than near the center. The twisted wing tips serve more or less as a substitute for the tail. This application of the sweepback for securing static stability of the wing is quite different from its use for delaying transonic difficulties, mentioned in Chapter IV.

Dynamic Stability

The complete problem of stability of an airplane is much more complicated than the foregoing remarks might indicate, because there is not only the question of static stability but also the more difficult one of *dynamic stability*. The difference between static and dynamic stability is best illustrated by an example. A top at rest is obviously statically unstable in its upright position, but if it is spun it certainly has a kind of stability. Another example of dynamic stability which everybody knows is the bicycle. How shall we define this kind of stability? Assume that a steady motion of a body, such as uniform rotation or rectilinear uniform translation, is disturbed by a small amount. We call the body dynamically stable if its subsequent motion remains within a certain neighborhood of the original undisturbed motion. For example, if we deflect the axis of the spinning top, the gyroscopic force stabilizes the motion, so that the upper end of the top describes a small circle or a system of cycloids in the neighborhood of its original position. A dynamically stable body does not necessarily return to its original state of motion. But it is necessary that the deviation from the original motion remain small provided the initial disturbance has been small. Evidently, without spin the top would fall, so that its upper end would continuously and rapidly move away from its original position.

The mathematical theory of dynamical stability was first formulated by the British mathematician, Edward J. Routh, in a book published in 1877 (Ref. 2). The theory was first applied to the stability of airplanes by Bryan and Williams in 1904 (Ref. 3). In the same year General Crocco, then a young lieutenant, published a paper (Ref. 4) on the stability of dirigibles. In this paper he arrived at the remarkable result that the horizontal flight of a dirigible can be dynamically stable when the ship is statically unstable. In other words, it is possible that a dirigible model put into a wind tunnel might show an unstable moment tending to increase an initial angular deviation and that, nevertheless, if one takes into account all the aerodynamic forces occurring in flight, the airship might be dynamically stable. The practical consequence of this result is that the size of the tail surfaces necessary for stable flight is significantly smaller than static stability would require.

Let us return to the general problem of the stability of an airplane. We consider the airplane as a rigid body with six degrees of freedom—three components of linear displacement and three of angular displacement. In the latter part of this chapter we will consider the airplane as an elastic system, taking into account the deformation of its wing and control surfaces, but here we imagine it to be rigid. We use a co-ordinate system in which the origin coincides with the center of gravity of the airplane (Fig. 61). The x and z axes lie in the symmetry plane, and the y axis is perpendicular to it. The x axis is in the direction of flight; the z axis has been decided by higher authorities to be measured positive downward, although I would prefer it positive upward. Components of the velocity of the center of gravity in the directions of the co-ordinate axes are denoted by u, v, and w. We call v a *sideslip* and w a *plunge*, although the latter may not be a common expression. Components of angular displacement about the co-ordinate axes are denoted by φ, θ, and ψ; they are called *roll*, *pitch*, and *yaw*, respectively. The positive direction of any angular displacement is determined by the rule that it is clock-

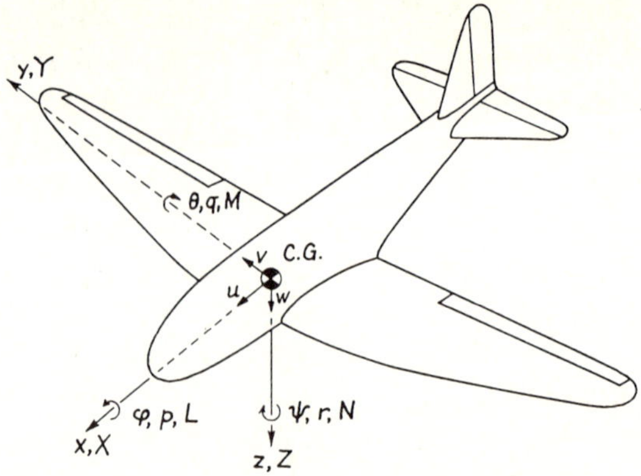

Fig. 61. Standard conventions for airplane stability discussions.

wise if one looks in the positive direction of the axis of rotation. The corresponding angular velocities are rolling velocity, *p;* pitching velocity, *q;* and yawing velocity, *r;* respectively. The linear forces acting in the directions of the co-ordinate axes are *X*, *Y*, and *Z;* the corresponding moments about these axes are *L*, *M*, and *N*, and are called *rolling*, *pitching*, and *yawing moments*, respectively. The control surfaces used in general to produce these moments are *ailerons* for rolling, *elevators* for pitching, and *rudders* for yawing. In the earliest airplanes warping of the wing surface (*gauchissement*) was used instead of ailerons. The *spoiler*, usually a kind of flap arranged on the wing's upper surface or emerging from a slot in the wing, "spoils" the circulation and therefore the lift. Spoilers applied alternately to the two half-wings can replace the ailerons. Sometimes ailerons and elevators are combined, especially in tailless airplanes. The combination is called *elevon*, a term created at Northrop Aircraft, Inc. Elevons work as an elevator when moved in the same sense, and as ailerons when moved in the opposite sense.

Now the question is how to deal with all these motions. Complications arise from the fact that the six degrees of freedom are not independent; certain motions are coupled. Suppose, for example, that a plunging motion is given to an airplane originally in steady horizontal flight. Then the velocity of the air relative to the airplane becomes inclined, i.e., the angle of attack is changed. The change of the angle of attack produces a pitching moment which sets up a pitching motion. One sees that there is a coupling between plunging and pitching. We call a motion in which every point of the airplane moves in a plane parallel to the plane of symmetry a *longitudinal motion* and the corresponding stability *longitudinal stability*. The longitudinal motion combines the motion of the center of gravity in the plane of symmetry with the pitching of the airplane about the center of gravity. Because of the symmetry in the airplane's geometry, the problem of longitudinal stability can be separated from that of lateral stability, which comprises the motions of roll, yaw, and sideslip.

Longitudinal Stability

As mentioned before, the static stability of a conventional airplane is maintained by its tail. Although static stability is concerned only with stability of equilibrium, it nevertheless plays an important role since it can be shown that—as far as longitudinal stability is concerned—practically all dynamically stable airplanes are statically stable.

The analysis of the motion of stable airplanes shows two distinct types of longitudinal motion; one is a slow motion of long period and the other a rapid motion of much shorter period. The former involves deviations from a straight trajectory of the center of gravity; the velocity of the center of gravity increases while the plane is diving and decreases while it is climbing. The angle of the wing relative to the trajectory is maintained almost constant. This type of motion was first described by Joukowski in 1891 (Ref. 5), and later independently by Lanchester (Ref. 6), both of whom I have already mentioned in connection with the theory of lift. Joukowski's contribution has been altogether overlooked, and the phenomenon is usually known as Lanchester's *phugoid* motion. My Greek would suggest "phygoid," but I am not quite sure about the term. At any rate, this strange word seems to come from Lanchester's misinterpretation. The Greek word φευγειν literally means "to fly" in the sense of fleeing before a menace and not flying as a bird. The various trajectories of the phugoid motion are reproduced in Fig. 62. According to the

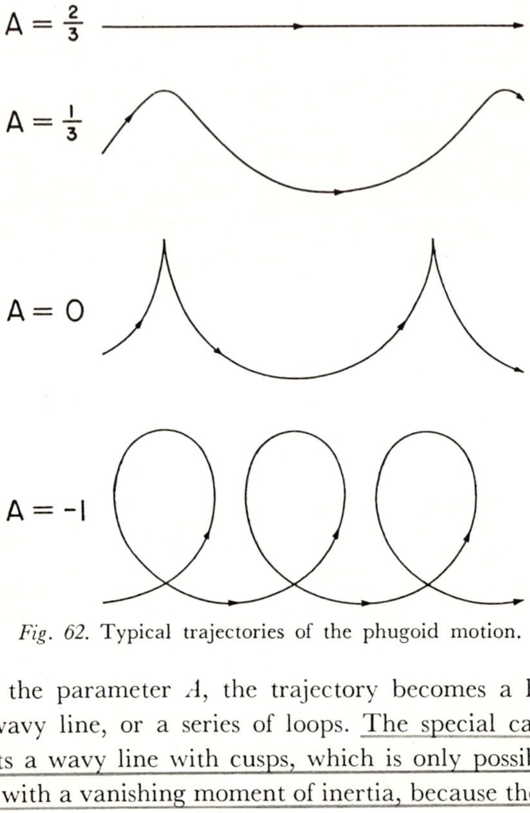

Fig. 62. Typical trajectories of the phugoid motion.

value of the parameter A, the trajectory becomes a horizontal line, a wavy line, or a series of loops. The special case $A = 0$ represents a wavy line with cusps, which is only possible for an airplane with a vanishing moment of inertia, because the airplane is required to turn 180° in no time at the cusp. Although the phugoid is an idealized type of motion, it still gives a reasonably correct picture of motions where the angle between the wing and the trajectory is kept constant. The general motion can be thought of as a pitching oscillation of short period superposed upon the phugoid oscillation. As a matter of fact, we do not often notice these phugoid oscillations in modern airplanes; their period is so long that they are either corrected by the pilot or obscured by gusts in bumpy weather.

Even if we travel over the weather, as modern airplanes do, we still do not often notice anything like a phugoid motion. What we do notice sometimes is the short-period oscillation. Pitching oscillations of short period are usually very quickly damped out, because the tail not only provides static stability but also damping in pitch. Insufficient damping is not pleasant for the passenger and makes the job of the gunner in military airplanes very difficult.

Difficulties are encountered when the airplane flies in the transonic region or at high angles of attack. I have already mentioned in Chapter IV the transonic troubles caused by sudden changes of pitching moment and the like. One difficulty occurring at high angles of attack is the so-called *buffeting*, usually caused by some vortex shedding, which may originate, for example, in the junction of wing and fuselage. Separation of flow may be caused because the junction forms a kind of diffuser—a tube of increasing cross section. Since the separation often occurs periodically, owing to vortex shedding, it may cause annoying oscillations. The trouble may be cured by a smooth fairing called a *fillet* between the wing and fuselage. This device was developed at the California Institute of Technology (Ref. 7) and first employed on the Northrop Alpha airplane.

This is a typical example of developments made in the wind tunnel and applied with success in practice. I worked on this problem together with Clark Millikan and Arthur Klein. In 1932 I gave a lecture in Paris on up-to-date problems in aerodynamics and mentioned the wing-fuselage fillet as an effective means of preventing buffeting. It turned out that at that time the French designers had the same trouble we had in the United States. One of the prominent designers told me later that after my lecture he tried a fillet right away on his new prototype and had success. Thus in France the fillet was connected with my name and was called a "karman." The French say an airplane has a "big karman" or a "small karman." I discovered this fact many years later on the occasion of a trip to France; people in aeronautical circles who heard my name asked: "The man with the fillet?" The invention of the wing-fuselage fillet was in reality a joint work of our C.I.T. team.

Lateral Stability

A few words about the other motions—sideslip, rolling, and yawing—are in order. These motions are coupled with each other. For example, if a yawing motion is given to an airplane originally in steady, straight flight, so that the left wing moves forward and the right wing backward, then the relative air velocity increases on the left wing and decreases on the right wing. This results in an increase in lift on the left wing and a decrease in lift on the right wing, thus producing a rolling moment on the airplane. On the other hand, if a rolling motion is given to the airplane, a yawing moment will be produced which tends to move the descending wing forward. In this way the rolling and yawing motions are coupled with each other. There are also other couplings between the motions, so that they must be considered together under the heading *lateral stability.*

The vertical fin furnishes static stability in yaw, also called *directional stability.* It is very difficult to get sufficient directional stability without such a surface somewhere, either on the wing or on the tail. There is no static stability in roll, because there is no rolling moment to right a banking airplane. What happens is that the horizontal component of the inclined lift represents a side force so that the airplane sideslips (Fig. 63). This coupling of sideslip and roll makes possible the achievement of dynamic stability in roll. It is achieved by giving to the two wing halves a flat V shape called *lateral dihedral.* The dihedral produces a rolling moment which tends to restore the airplane to its normal flying position. This effect was known to early investigators, for example, to Sir George Cayley. The function of the dihedral

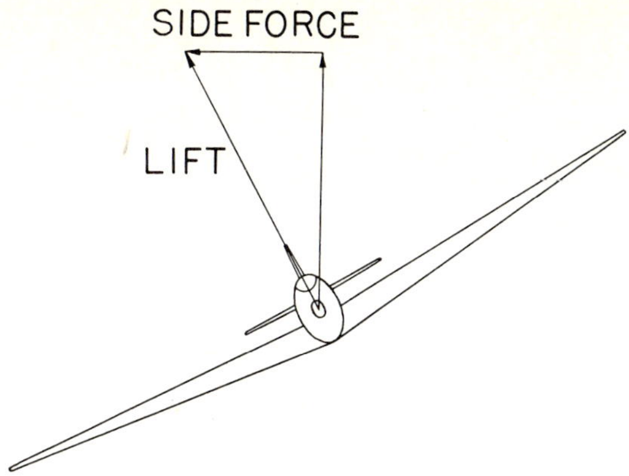

Fig. 63. Production of side force on an airplane in a bank.

can be explained in the following way: Let us consider, for simplicity, a wing of rectangular form without sweepback. In Fig. 64, *A* is a point on the leading edge of the right wing, *AB* and *BC* are the components of velocity in the x and y directions, respectively, and the plane *ADE* is the plane of the chord of the right wing. If there is no sideslip, *C* coincides with *B*, and the angle *DAB* is the angle of attack. In the presence of a sideslip, *BC*, the angle of attack is represented by the angle *EAC*, which is greater than the original angle *DAB*. The opposite is true for

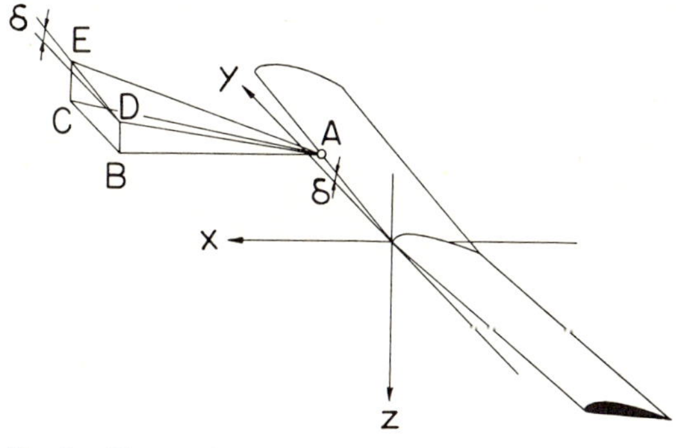

Fig. 64. The explanation of the function of dihedral when sideslip occurs. The angle δ is the dihedral angle.

points on the left wing. Therefore the wing gets more lift on the side toward which it is slipping and loses lift on the opposite side, thus creating a rolling moment to restore the wing to its original position. There is a similar effect due to sweepback of the wing, but the rolling moment due to sweepback is proportional not only to the sweepback but also to the angle of attack, whereas the rolling moment due to dihedral is proportional to the dihedral angle and is independent of the angle of attack.

Now the lateral stability of an airplane is achieved by compromising the requirements of the static directional stability due to the vertical fin and the dynamic stability due to the dihedral. If the dihedral effect is too strong, the airplane, when banked, rolls back too far, so that it sideslips the other way and overshoots again, thus executing a motion called a "Dutch roll." (The name probably came from its resemblance to a skating figure that the Dutch people used to perform.) This sort of motion is not a real instability but is unpleasant and undesirable. It is actually harmful in military airplanes, where accurate shooting is required. The airplane designer is usually ashamed if his airplane dances in this way. The other extremity, which occurs when the directional stability is exaggerated, is a real instability. If the airplane gets a small initial yaw to the right, for example, this is followed by a bank to the right. This bank causes additional yaw, which is followed by more bank, and the process continues. The motion starts as a gentle spiral, which, if left to itself, gets continually tighter and steeper. For this reason, the motion is called *spiral instability*. Unfortunately the motion develops so gradually that the pilot is often unaware that his airplane is deviating from straight flight. In the air it is rather difficult to know whether the flight path is straight or is a circle of large radius, without reference to the ground or to some other fixed direction such as may be given by the stars or a gyroscopic instrument; the increase of the resultant gravity due to the centrifugal force in the case of flight in a large circle is so small that the pilot cannot sense it.

Most airplanes have a certain degree of spiral instability. If the instability is reasonably weak, it is usually corrected by the pilot; only excessive instability should be avoided. Theoretically speaking, most airplanes are not completely dynamically stable. In other words they cannot be flown "hands off" indefinitely.

In order to be more precise in statements about the stability of an airplane, two aspects of the subject must be introduced that were not mentioned before. First, the effect of an initial disturbance depends essentially on whether or not the control surfaces deflect during the subsequent motion. It is evident that two extreme assumptions can be made, namely, that the controls are fixed in their initial positions and that they are completely free to move on their hinges. The first assumption corresponds very closely to the case of an airplane with power-actuated control surfaces, which are usually irreversible in the sense that the aerodynamic forces cannot cause them to deflect against the control mechanism. The second limiting case, controls free, is a somewhat idealized representation of an airplane with manually operated controls, flown by a pilot who allows the airplane to fly "hands-off." The degree of stability in these two limiting cases may be quite different, so that the certainly desirable goals of both control-fixed and control-free stability may sometimes be difficult to attain.

The second aspect of the stability problem to which no reference was made earlier is the influence of the propulsion system. One must consider both power-on and power-off stability. The difference is mainly due to two factors: one is the direct effect of the thrust on the equilibrium and motion of the airplane; the second is change of aerodynamic forces acting on wing and tail because of the flow induced by the propulsive system. The latter effect is generally more significant in propeller-driven airplanes than in those using jet engines; it is called the slipstream effect. Even in jet airplanes, most designers locate the tail surfaces well above the jet stream in order to avoid interference.

Lateral Motions above the Stall

The stability or instability we have discussed heretofore is concerned with the wing working below the stalling angle. In this range the lift increases with an increase in the angle of attack. If the angle of attack exceeds the stalling angle, as mentioned in Chapter II, the lift decreases with increasing angle of attack. This makes possible a phenomenon known as *autorotation.*

Consider a wing with rolling velocity superposed upon straight flight velocity. The relative airflow has a larger angle of inclination for the wing moving downward and a smaller angle for that moving upward. Below the stall the lift is largely proportional to the angle of attack, so that the lift on the wing moving downward is increased while that on the other wing is decreased. The result is a rolling moment which damps out the rolling motion. Above the stall, however, a larger angle produces a smaller lift, so that a rolling moment is set up which accelerates the initial rolling instead of retarding it. The resulting steady rotation is known as autorotation of the wing.

The stall usually does not occur simultaneously at all sections along the span. If the central portion stalls first while the tips remain unstalled, the damping contributed by the tips is usually sufficient to counteract the negative damping due to the central portion. Furthermore, in such a case the ailerons still retain their effectiveness. On the other hand, if the tips stall before the center, the damping in roll disappears and also the ailerons lose effectiveness. If such a stall occurs near the ground, recovery may be almost impossible and a serious accident may result. The designer must avoid tip stall by reducing the angle of attack of the tip portion or by judicious variation of airfoil sections along the span. The tip stall is accentuated in the case of wings with a high taper ratio and especially in those with large sweepback. For such wings it is frequently necessary to provide high-lift devices, such as leading-edge slots, for the portion near the tip.

The motion of the airplane caused by autorotation of the wing is known as *spin.* The airplane descends along a helical path while it is continuously rolling and yawing. The best way to recover from a spin is to decrease the angle of attack; the airplane then goes into a normal dive. In many cases, however, the distribution of masses in the airplane is such that a gyroscopic moment tends to increase the angle of attack. Consequently, very large control forces are necessary to recover from the spin. The ailerons are almost always useless, and the elevator often loses its effectiveness; hence the rudder is frequently the only control surface that remains useful. It is therefore advisable to design the tail surfaces in such a way that the rudder will not be shielded by the horizontal tail in a spin. As a matter of fact, the spin is not an instability and is not always dangerous; evidently some pilots like to spin. Not all airplanes are able to spin. It depends on the stalling characteristics and the distribution of masses.

I once met the famous British aviatrix Amy Johnson at a so-called *conversazione* of the Royal Aeronautical Society, where the problem of spin was discussed by British and American engineers and scientists.

She came to me and asked, "Can you tell me in a few words what causes a spin and what is the mechanism of the thing?"

"Young lady," I told her, "a spin is like a love affair; you don't notice how you get into it and it is very hard to get out of!"

Aeroelasticity

In the theory discussed above we have assumed that the structure of the airplane is rigid. This assumption is justified so long as the stiffness of the structure is great and the speed of flight is small, but the effects of deformation of the structure cannot always be neglected, especially at high speeds. Such effects are covered by the term *aeroelasticity.* Aeroelasticity deals with the mutual influence between aerodynamic forces and elastic deformations.

Consider the wing of the airplane as a beam. A beam has a so-called elastic axis; if the lift force acts on this axis, the result is a simple bending without accompanying torsion. But if the lift acts ahead of the elastic axis, the resulting deformation is a bending plus a torsion, the latter tending to increase the angle of attack. This in turn increases the lift and therefore the torsion. Of course, the elastic stiffness of the wing resists this deformation. However, since the aerodynamic force increases approximately with the square of the flight speed, while the elastic stiffness is independent of the speed, there must be, theoretically, a critical speed at which the two effects are equal and above which elastic instability occurs. This speed is called the *divergence speed*. It is seldom encountered in actual flight; I observed it once in my life—it was a very sad experience. In 1922 a sailplane called *Weltensegler*—the "World Sailor"—entered a sailing contest in the Rhön mountains. It was built by a group of ambitious students who apparently did not know enough about elasticity and aerodynamics; it had an aspect ratio of more than 20. At first the pilot sailed successfully in an updraft. However, when he came out of the updraft region he went into a dive with increasing speed. As we watched from the top of the mountain, the wing of the glider was slowly twisted off!

Another trouble due to elastic deformation is the *reversal of control*. Consider, for example, a conventional aileron. If the wing structure is rigid, a downward deflection of the aileron produces an increase of the lift and therefore a rolling moment which tends to raise the wing tip. But if the wing structure is flexible, the wing twist caused by the aileron deflection diminishes the angle of attack at the wing tip and thereby reduces the lift acting on the tip section and the rolling moment. Thus, the actual rolling moment may be essentially smaller than the same aileron deflection would produce on a rigid wing. In other words, the aileron loses some of its effectiveness. Since this effect increases with flying speed, there will be a critical speed at which the aileron is completely ineffective, and for still higher speeds the action of the aileron will be reversed.

If we take the elastic effects into account, wing theory becomes more complicated than it appeared in Chapter II. For a rigid wing, the effective angle of attack of the relative wind at any cross section, which determines the lift and drag of the section, is obtained by combination of the flight velocity and the induced downwash. For an elastic wing, the magnitude and direction of the relative wind depend also on the elastic deformation, which is in turn influenced by the same lift distribution that we endeavor to calculate. Sears has proposed an approximate method for the calculation of this mutual interaction (Ref. 8). Aeroelastic effects are important for all high-speed airplanes. If the aspect ratio is large, the wing twist is significant. For airplanes with small aspect ratios, we encounter some other types of aeroelastic deformation, such as chordwise bending.

Finally, we should consider the combined effects of elastic and inertia forces. One simple example is the following: Assume that a sweptback, elastic wing performs a plunging motion. The increase of incidence due to the plunging tends to bend the wing tips upward. But since the plunging is decelerated by the increased lift, the inertial forces tend to bend the tips downward. This is an example in which essential difference exists between real flight and its simulation in a wind tunnel; in the wind tunnel the motion of the model is usually restricted, so that the elastic forces are simulated but without the compensating inertial forces.

The most important example of the co-operation of aerodynamic, elastic, and inertial forces is what we call *flutter*. I will sketch here the simplest case. Consider a wing fitted with a hinged control surface and assume that the wing performs a bending oscillation in an airstream. The frequency of this oscillation is essentially equal to the elastic frequency of the wing; it is somewhat influenced by the speed of flight, but the effect is small. We assume for simplicity that the control surface is completely free. Since it is exposed to the airstream, it becomes effectively stiff, just like a weather vane; it has an apparent elastic stiffness. This apparent stiffness determines the frequency of the oscillation of the control surface; its frequency evidently increases with the speed of the airstream. If its frequency coincides with the bending frequency of the wing, one observes a large increase of the amplitude of the oscillations.

In this simple case the flutter has the character of a *resonance*. Perhaps the simplest example of a resonance is that of a pendulum whose point of support is kept in oscillating motion with a frequency equal to that of the pendulum. It is easy to show experimentally that in this case the pendulum will go into large oscillations. The phenomenon of resonance is cleverly used by people predicting hidden processes with a pendulum. They predict, for example, the existence of water or ore beneath the ground. They tune the pendulum to the frequency of their pulse, so that the slightest movement of their hand causes the pendulum to oscillate with considerable amplitude. Our simple flutter case is based on a similar principle.

The wing, being elastic, always oscillates slightly, so that the hinge of the control surface is in periodic motion, even if this is not visible to the naked eye. This motion is not objectionable, except when the frequency of the control surface becomes equal to the frequency of the wing. In this case resonance results and both the wing and the control surface develop large amplitudes of oscillation. The reader may wonder what is the source of the relatively large kinetic energy of this violent oscillation. It is true that the relative airstream tends to damp the bending vibrations of the wing, but the oscillations of the control surface take energy out of the airstream and excite the oscillation of the wing instead of damping it. This example is somewhat oversimplified, but it serves well to show how *self-excited oscillations* can exist at a certain speed or in a certain speed range. Actual flutter phenomena are much more complicated; for example, resonances are possible between any combinations of bending and torsional oscillations of the wing and many types of oscillations of the control surface. Flutter is an important and difficult problem of aeroelasticity; many aeronautical engineers specialize in it. Every major aircraft company has a department devoted especially to the flutter problem.

Some years ago, when 450 to 500 miles per hour was still a high speed, the president of an aircraft company in California got a telephone call from Wright Field that his prototype airplane had encountered serious flutter at 450 miles per hour.

The president called in his vice-president in charge of engineering and said, "This is a scandal! We have the best mathematicians, a whole department for flutter, and still General X calls me from Wright Field to say that we have flutter at 450 miles per hour!"

So the vice-president went to the head of the flutter group and said, "They have telephoned from Wright Field that our new airplane has flutter at 450 miles per hour!"

To which the engineer replied, "Oh really? I am glad to hear that. In my report to you I predicted flutter at 445 miles per hour!"

The science of aeroelasticity, including flutter theory, is now in a period of active development. This is especially true because the large forces acting on airplane parts in high-speed flight require the designer to analyze more and more exactly the elastic deformations of the airplane structure. Although the mathematics of the problem have become more complex, the development of novel computing devices allows the engineer to obtain the solution of complicated systems of equations in much shorter time than before and with increased accuracy. Not only have such computing devices found wide application in aeronautical engineering, but some aeronautical engineering organizations have even gone out of their way to improve and produce new computing machines.

References

1. Pénaud, A., "Aéroplane automoteur," *L'Aéronaute*, 5 (1872), 2–9. Partially reprinted in F. W. Lanchester's *Aerodonetics* (London, 1908) Appendix I.

2. Routh, E. J., *A Treatise on the Stability of a Given State of Motion, Particularly Steady Motion* (London, 1877).

3. Bryan, G. H., and Williams, W. E., "The Longitudinal Stability of Aerial Gliders," *Proceedings of the Royal Society of London*, series A, *73* (1904), 100–116; Bryan, G. H., *Stability in Aviation* (London, 1911).

4. Crocco, G. A., "Sur la stabilité des dirigeables," *Comptes rendus de l'Académie des Sciences, Paris, 139* (1904), 1195–1198; "Sulla stabilità dei dirigibili," *Rendiconti della Reale Accademia dei Lincei, classe di scienze fisiche, matematiche e naturali, 13* (1904), 427–432.

5. Joukowski, N., "On the Flight of Birds" (in Russian), *Obshchestvo liubitelei estestvoznaniia, antropologii i etnografii, Moskva, Izviestiia, 73* (1891), 29–43.

6. Lanchester, F. W., *Aerodonetics* (London, 1908).

7. Kármán, Th. von, "Quelques problèmes actuels de l'aérodynamique," *Journées techniques internationales de l'aéronautique* (*1932*), 1–26 (Paris, 1933); Klein, A. L., "Effect of Fillets on Wing-Fuselage Interference," *Transactions of the American Society of Mechanical Engineers, 56* (1934), 1–7 (AER–56–1).

8. Sears, W. R., "A New Treatment of the Lifting-Line Wing Theory, with Application to Rigid and Elastic Wings," *Quarterly of Applied Mathematics, 6* (1948), 239–255; Pai, S. I., and Sears, W. R., "Aeroelastic Properties of Swept Wings," *Journal of the Aeronautical Sciences, 16* (1949), 105–115.

CHAPTER VI

» *From the Propeller to the Space Rocket*

MANKIND'S yearning to fly was the force which drove inventors and scientists to find out how to fly. Another psychological trend that is even more general and even older than flight, but most evident in the history of flight during the last fifty years, is mankind's yearning for speed. We often hear of new speed records but seldom hear anyone question why it is necessary to travel so fast. Who knows whether the world would not be happier without the great speeds at which we now move? But it seems that striving for greater speeds is human nature. While economic reasons may influence this desire for greater speed to some extent, the principal motivating factor seems to be psychological—perhaps just the love of setting new records. Young students with athletic ability may have brilliant scientific minds and still believe that to jump two inches farther than anybody else is an important contribution to human progress!

What Price Speed?

Before I start on the subject of this chapter, namely, the history of our knowledge of aerial propulsion, I want to mention a study that I made a few years ago, partly for fun, partly from scientific interest (Ref. 1). Together with Guiseppe Gabrielli, the well-known airplane designer and director of the Fiat airplane factories, I made a kind of factual—not theoretical—survey of existing vehicles, including those moving on the earth, on and in the water, and in the air, from the viewpoint of how much power they have available per unit weight. The study was called "What Price Speed?" For the purpose of comparison we plotted the specific power, defined as the ratio of the maximum power available to the gross weight of the vehicle, as a function of its maximum speed (Fig. 65). We found a kind of limiting line, a straight line in the logarithmic diagram, such that all presently known

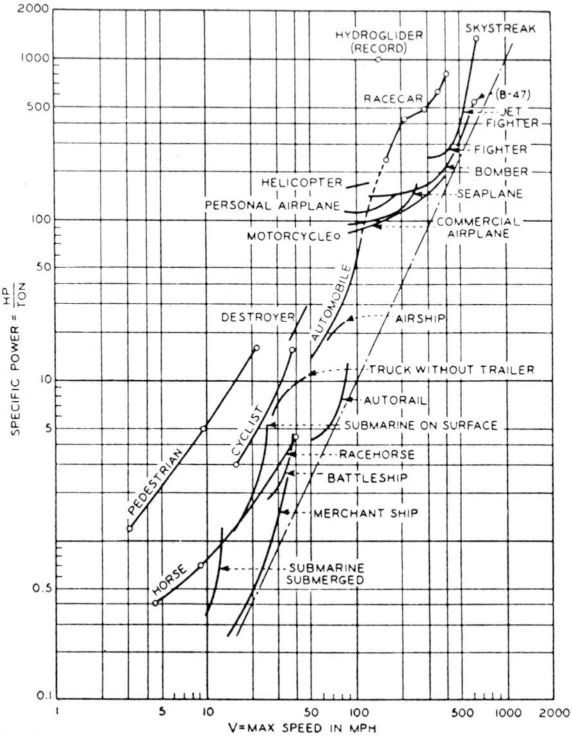

Fig. 65. Specific power, defined as the ratio of the maximum power available to the gross weight of the vehicle, plotted as a function of its maximum speed. (From G. Gabrielli and Th. von Kármán, in *Mechanical Engineering*, *72* [1950], 776, by permission of the American Society of Mechanical Engineers.)

single vehicles are on its left side. (As "vehicles" we included also the pedestrian, the horse, the cyclist,—but no fish or fowl.) If any point is far above the straight line, it means that it is not as economical as it could be for the same maximum speed. For example, we see how economical merchant ships are until they come to a certain speed, at which point the specific power goes up suddenly. The data presented in the diagram are not average values but best values; hence the power indicated by each curve represents the minimum value necessary for the given type of vehicle. The pedestrian, cyclist, and horse are estimated from some scientific calculations. I do not have any explanation why the race horse, whose speed is about 40 miles per hour, has exactly the same specific power as a good battleship!

This diagram has still another interesting characteristic. If the specific power is proportional to the speed, the total work necessary for transportation over a given distance is the same. This condition corresponds to straight lines of 45° slope in the logarithmic diagram. We can therefore say that any vehicle performs best where its curve has a 45° slope. If the slope is less than 45°, the vehicle is improved by increasing its speed. If the slope is greater than 45°, this is a sign that the vehicle is beyond its best application. For example, if we take the curve for commercial airplanes, we see that between 200 and 300 miles per hour the slope is about 45°, or a little less. It is really true that the faster Constellation is more economical than the slower DC–3, if economy is measured by the horsepower-hours necessary to transport a load over a given distance.

Jet fighters have a curve of steeper slope; they cannot, of course, be called economical. There are two reasons for using less economical transportation; one is that with greater speed you can use your vehicle more. The number of hours per month being the same, the distance flown will become greater. This is the philosophy under which jet-driven airplanes are being promoted for commercial use. They may permit a greater number of passenger-miles per year, assuming that the airlines can fly them the same number of hours; i.e., that they do not require more time for maintenance. Secondly, of course, the use of noneconomical vehicles may be imperative for military purposes, where it is necessary to be faster than the other fellow.

I would conclude that if we want to arrive at a judgment concerning different methods of propulsion to secure speed, we have to take into account several points. First, the fundamental economy of the power requirement; second, the practical economy of its use in transportation; third, all the other viewpoints—psychological, political, and the like. What a human being will pay for speed is very hard to predict. How much will the average man pay for a trip of five hours which would otherwise take ten hours? I recommend this question to economists, psychologists, and other representatives of the social sciences as a worth-while study.

Theory of Propellers

The recent great progress in propulsion is generally known as the transition from the propeller to propulsion by reaction.

A few years ago I was in Paris during the time of peace negotiations between the Allies and some Eastern European countries. A Hungarian woman journalist came to me and wanted an interview. She asked me what I thought was the greatest progress in aviation in the last decade. I told her, "Propulsion by reaction."

She said to me, "Professor, could you express this in some other way? I cannot write in a progressive paper that progress is accomplished by reaction!"

I tried to find a Hungarian word for jet and she left, apparently satisfied.

In reality, differentiation between propulsion by propeller and propulsion by reaction is not quite correct. From the viewpoint of general principles of mechanics, the propeller is also a device for propulsion by reaction. This was not the opinion in early times, however, when propellers for lifting weight in air, and later for driving vehicles in water and air, were invented and investigated. The fundamental concept then was that the propeller blade is part of a helicoidal surface which penetrates the fluid medium as a screw penetrates a solid body. Consider a screw jack: the screw advances one helicoidal pitch with every revolution, and, if we neglect friction, the work necessary to turn the screw is equal to the work necessary to lift the weight. It was imagined that the propeller is analogous to a screw that penetrates a fluid medium—air or water. The fluid, however, yields, whereas a solid does not. Consequently, the advance of the propeller in the axial direction is less than the pitch of the helicoidal surface. Then the primitive propeller theory said that, if the advance in the axial direction were equal to the helicoidal pitch, the propeller would be 100 percent efficient, since in this case, neglecting friction, the work done by the advance is identical with the work required to turn the propeller. When the propeller slips, the thrust multiplied by the advance of the propeller represents the useful work done; therefore its efficiency is equal to 1 minus its percentage slip. This constitutes a primitive but intuitive kind of propeller theory.

The man who demonstrated that the functioning of the propeller is based on the principle of reaction was Rankine. He was a very ingenious engineer, mentioned earlier in Chapters III and IV. At a relatively early date when not all engineers appreciated the importance of basic knowledge, he was an advocate of research and the training of engineers in fundamental science. When somebody told him that a practical engineer does not need much science, he said, "Yes, what you call a practical engineer is the man who perpetuates the errors and mistakes of his predecessors." Although the definition is somewhat hard on many good practical engineers, it is correct in the sense that engineering education should not only transmit experience from generation to generation, but should be based at all times on the old and new developments in fundamental sciences.

Rankine (Ref. 2) recognized that the essential point in the action of a propeller is the acceleration of the air mass passing through the circular area swept by the propeller blade; we sometimes call this circular area a *propeller disk*. Let us assume that the propeller disk is stationary in an airstream. The air moves toward the disk with the velocity U (Fig. 66), and its velocity ultimately

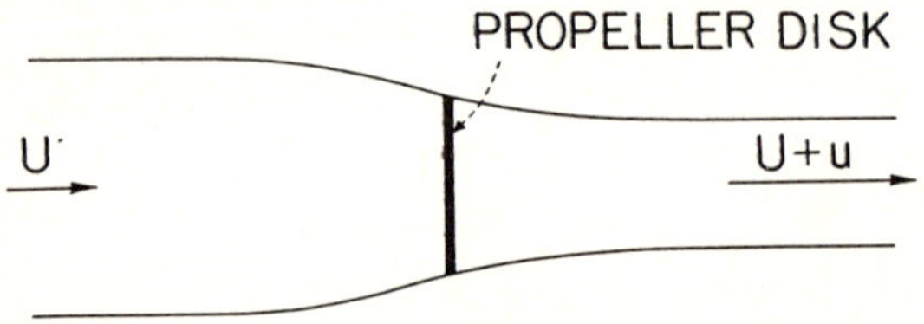

Fig. 66. A diagram illustrating the momentum theory of the propeller.

is increased by an amount which we denote by u. In other words, the propeller takes a mass of air—if used as a helicopter, from above; if used for propulsion, from the front—and accelerates it downward or backward, respectively. The rate of change of momentum is equal to the thrust. If Q is the mass of air which goes through the propeller in unit time, then the product Qu is the rate of change of momentum. On the other hand, if we consider the propeller advancing with the velocity U through air at rest, the work which has been expended is equal to the increase of kinetic energy of the air: $\frac{1}{2}Q[(U + u)^2 - U^2]$, or $Qu(U + \frac{1}{2}u)$. Now, if we define as efficiency η the ratio of the useful work done, QuU, to the total work expended, $Qu(U + \frac{1}{2}u)$, we get the following formula:

$$\eta = \frac{QuU}{Qu(U + \frac{1}{2}u)} = \frac{1}{1 + \dfrac{u}{2U}}.$$

This quantity is, of course, always smaller than 1. We call it the *propulsive efficiency*. In order to have good propulsive efficiency, i.e., a value near unity, the velocity increment, u, must be small in comparison to the flight velocity, U. For example, if u is equal to U, i.e., if the acceleration is 100 percent, the efficiency is only 67 percent. This calculation does not include all the losses, such as, for example, the loss due to rotating motion imparted to the air or the friction on the propeller blades.

The principle that an efficient propeller requires as small a value as possible for the velocity increment of air passing through it applies to other propulsive devices based on the reaction principle. We often have to tolerate jet velocities that are high compared to the flight velocity, although we know that the propulsive efficiency will be poor. For example, with rockets the outflow velocity of gas may be equal to 5,000 to 6,000 feet per second, whereas the flight velocity may be only around 900 feet per second. One can easily calculate how poor a rocket airplane would be for commercial purposes. Rankine's simple theory furnishes a result of prime importance.

We may ask the magnitude of the maximum thrust that can be developed by a propeller of given size, say by a propeller with a disk area equal to S. In order to calculate this value, we have to assume a relation between the air mass flow, Q, and the area, S. In general, one assumes that the average velocity of the air traversing the disk area is the arithmetic mean value between the velocity, U, far ahead, and the velocity $U + u$, far behind, the propeller. With this assumption one can show first that, provided the same work is expended per unit time, the maximum thrust is reached when $U = 0$, i.e., if the propeller stands still and the air is originally at rest. In this case the ratio of power required, P, to the thrust available, T, is given by

$$\frac{P}{T} = \frac{1}{\sqrt{2}} \sqrt{\frac{T}{\rho S}},$$

where ρ is the density of air. This formula applies, for example, to a hovering helicopter, for which T is equal to the weight, W, to be supported. We remember that a similar formula was obtained by early researchers for the work required for sustentation of a weight, W, by an airplane wing (see Chapter I). The numerical factor is different, to the disadvantage of the helicopter.

However, an airplane is unable to hover, and therefore the larger power requirement for the helicopter is acceptable.

Theory of Propellers: Connection with Wing Theory

The so-called *momentum theory* of the propeller initiated by Rankine and sketched above is based on the changes in momentum and kinetic energy of the airstream passing through the propeller disk. The rate of change of the momentum determines the thrust, but the theory does not tell anything about the way in which the thrust is transmitted from the air to the propeller structure. Similarly, Rankine's theory states that an amount of power equal to the rate of increase of the kinetic energy of the stream has to be put in by the rotating propeller, but it does not make any statement as to how the work expended by the torque is transferred to the air. On the other hand, the *blade-element theory* is based on the opposite concept; it considers the propeller blades moving through the air and computes the forces transferred from the blades to the air.

Fig. 67 shows schematically a section of the blade. Suppose

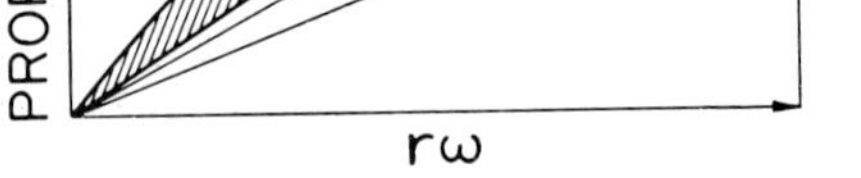

Fig. 67. The velocity relative to the blade element of a propeller. U denotes the forward speed, w the angular velocity, and r the radius of the element.

that the distance of the blade section from the axis of rotation is r, the angular velocity of the propeller ω, and the velocity of advance (i.e., the flight speed) U. Then, in first approximation, $r\omega$ and U are the components of the relative velocity between solid and fluid. Thus, if we consider the blade section as a wing section, lift and drag acting on the section can be computed, and by resolving the resultant force into components in the axial and tangential directions, we obtain the contribution of the blade element to thrust and torque. Summing up the contributions from all blade elements, we can obtain the total thrust and torque.

The two theories consider the same process from two entirely different points of view. The momentum theory is based on application of the basic laws of mechanics to a system comprising a fluid stream and a body moving relatively to it. On the other hand, the blade-element theory is based on our knowledge or assumption concerning the local interaction between fluid and solid. These two methods go parallel through almost the whole field of fluid mechanics; scientists and engineers are satisfied only if they can convince themselves that both methods lead to the same result. For propellers a satisfactory solution was obtained by combination of the two theories.

The blade-element theory was initiated by William Froude (Ref. 3), the famous English engineer whom we mentioned in connection with the problem of skin friction. Some years afterward the same theory was independently worked out in detail by Stefan Drzewiecki (1844–1938) (Ref. 4), an engineer and scientist of Polish origin, one of the most distinguished pupils of Joukowski. Drzewiecki later lived in France and worked with Eiffel. I had the pleasure of meeting him in Paris. I remember that at the age of seventy-seven he drove his own car all over France, going from airport to airport to watch the air races.

Drzewiecki recognized that it may be incorrect to apply the lift and drag coefficients of a wing of infinite span to a blade element like that of Fig. 67; he made a correction, assuming an equivalent aspect ratio for each blade. However, a logical solution of the problem could be reached only by a combination of the blade-element theory and the momentum theory. The problem is similar to that which occurs in the theory of wings with finite aspect ratio, viz., one has to determine the effective relative velocity between fluid and wing section in magnitude and direction.

The momentum theory clearly indicates that the axial velocity of the stream passing through the propeller disk is higher than that far ahead of the propeller. On the other hand, the rotating blades create a rotation in the air, so that the moment of momentum of the rotating air must be equal to the torque acting on the propeller shaft. Therefore the air in the plane of the propeller rotates in the same direction as the propeller blades. Thus the relative velocity in the tangential direction is smaller than $r\omega$. The diagram in Fig. 67 indicates the correction which one has to apply to obtain the appropriate values of the components of the relative velocity. With these components of relative velocity, we can determine the corrections of the lift and drag acting on the blade element in direction and magnitude.

Evidently, the corrections in velocity components represent induced velocities; the progress, compared to the primitive blade-element theory of Froude, is analogous to the progress of the wing theory achieved by Prandtl. Concerning the determination of the induced velocities, two steps can be recognized in the development of the theory. The first step was the combination of the blade-element theory and the momentum theory. The momentum theory makes it possible to compute mean values of the induced velocities. This method is identical with the assumption that the actual blades are replaced by a large number of uniformly distributed blades. It furnishes very satisfactory results, especially if one applies a correction proposed by Prandtl (Ref. 5) for the effects of blade tips. This refinement takes into account the influence of the number of blades.

The propeller theory sketched on these lines was worked out in the period 1918 to 1924 by Betz (Ref. 6) and Helmbold (Ref. 7) in Germany, by Wood (Ref. 8) and Glauert (Ref. 8) in England, and by Pistolesi (Ref. 9) in Italy. I might also mention a paper published jointly by Theodore Bienen and myself in 1924 (Ref. 10).

The second step in the development constitutes a direct application of the Lanchester-Prandtl ideas to rotating bound vortices representing the propeller blades. Helicoidal vortex sheets now replace the free vortex sheets of Prandtl's theory. This idea was first carried out mathematically by Sydney Goldstein in his doctor's thesis at Göttingen University (Ref. 11). Goldstein became one of the leading aerodynamicists in England and organized a group working on fluid mechanics in Manchester. At present he is active at the Institute of Technology in Haifa. Two Japanese aerodynamicists, Moriya (Ref. 12) and Kawada (Ref. 13), continued the work of Goldstein.

It is gratifying to see the progressive clarification of ideas on the functioning of a simple device like a propeller, from the analogy with a screw jack to a complete theory based on the principles of scientific fluid mechanics and using all the mathematical methods of this science.

From a practical point of view, great progress has also been made in the construction of propellers. I want to mention automatic pitch control and thrust reversal, the latter used by modern airplanes for braking effect. Sometimes a propeller may go into reverse thrust when it is not supposed to; the design does not seem to be absolutely perfect yet. The latest progress concerns propellers for very high speeds, e.g., supersonic speeds. The difficulty here is that, as we have seen in Chapter IV, the drag at supersonic speeds depends to a large extent on the thickness of the wing section. A supersonic propeller must therefore have very thin blades, which, however, introduce difficulties of possible vibration and excessive deformation. Thus the design of such propellers and the finding of appropriate materials and blade shapes present serious problems.

Jet Engines and Rockets

For almost forty years from the date of the first powered flight, the propeller driven by a reciprocating internal combustion engine was the only means of aerial propulsion. Of course, in these years the aircraft reciprocating engine made tremendous progress. For example, we have mentioned that the engine used by the Wright brothers had a weight of 15 pounds per horsepower; this ratio has been reduced to less than one pound per horsepower. In addition, a system of new propulsion devices competing with the conventional engine and propeller is now either in use or development. Fundamentally, as we have already mentioned, all propelling devices are based on the reaction or jet principle. They differ mainly in the manner in which energy is utilized to activate the jet.

Power for aerial propulsion can be produced by using the oxygen of atmospheric air as a chemical reactant in combination with some fuel, e.g., a hydrocarbon such as gasoline or kerosene. A second class of propulsion devices uses propellant combinations that produce power without the use of atmospheric oxygen. Such propulsion devices are called rockets. Finally, nuclear reactions can be used as the source of power.

Let us consider in more detail these three classes of propulsion devices. The devices using air and fuel can be subdivided according to the method of activating the jet whose reaction furnishes the propulsive thrust. With the propeller, the jet is produced by purely mechanical means. The propeller was driven exclusively by reciprocating engines, i.e., piston engines, until the lightweight gas turbine was developed as a prime mover. The propeller and gas-turbine combination is called *turboprop*—not a pretty English word but almost generally accepted. The compound engine, which is also used to drive propellers, is a composite of a piston and a turbo-engine.

Jet propulsion proper is different from the propeller in that the jet is activated by the introduction of heat energy, e.g., by combustion of the fuel in the atmospheric air. Such devices are called thermal jet-propulsion engines. The basic idea of such an engine is to produce high-pressure, high-temperature gas, which, being ejected from a tail pipe, furnishes the thrust. In the beginning of the development it was an item of discussion whether a combination of reciprocating engine and compressor or a gas turbine should serve as gas generator. Present jet engines use the gas turbine exclusively. The jet-propulsion engine designed by Secondo Campini and flown in 1940 in the Campini-Caproni airplane used a reciprocating engine. However, the first jet-propelled airplane ever flown (1939), the German Heinkel 178, used an engine of the type called the *turbojet* (Fig. 68).

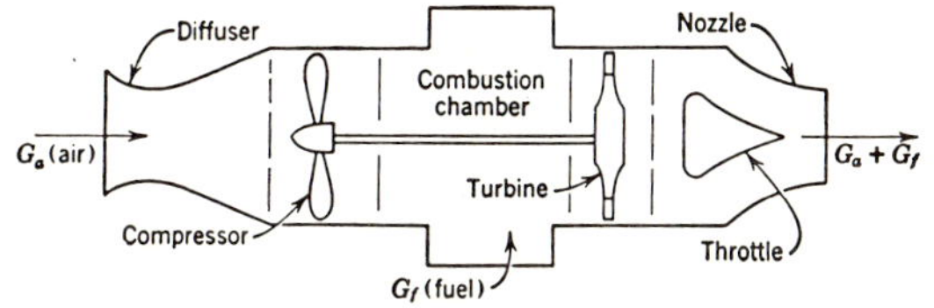

Fig. 68. Schematic diagram showing the elements of a turbojet. (From M. J. Zucrow, *Jet Propulsion and Gas Turbines* [copyright 1948, John Wiley and Sons, Inc.], by permission.)

The main parts of such a device are (*a*) a compressor, which takes in air from the outside and brings it to a certain pressure in order to make the combustion and the transformation of heat into mechanical energy more economical, (*b*) a combustion chamber or combustor, in which fuel is injected into the air-stream and burned, and (*c*) a turbine. The shaft output of the turbine drives the compressor, and ordinarily the gas leaves the turbine at high velocity and forms the jet that makes the thrust. We see that the turbine-compressor combination ultimately serves as a gas generator for producing the jet.

The first turbojet engine, mentioned above, the He S–3b, was designed by Hans-Joachim Pabst von Ohain, an engineer educated at Göttingen and employed by the Heinkel company. This engine developed about 1,100 pounds of thrust. Its compressor was of the centrifugal type and the turbine had radial inflow.

The development of jet engines in England and the United States is intimately associated with the work of Sir Frank Whittle. I do not want, however, to go into the details of this history. An excellent monograph by Robert Schlaifer (Ref. 14) gives a very complete account of the developments in the critical period preceding and during World War II, in various countries.

Some of the components of turbojets, like the centrifugal compressor and the turbine, had been used before as parts of conventional engines, viz., in superchargers for reciprocating engines in high-altitude flight. Combustion chambers were also known, of course, but the combustion of fuel in an airstream of relatively high velocity was a novel problem.

The thrust of the largest units built at present is of the order of 15,000 pounds. In many turbojets the thrust, at least for a short period, can be substantially increased by *afterburning*, i.e., by the injection of additional fuel into the tail pipe, using up the surplus oxygen contained in the jet. This is, however, an uneconomical process. Centrifugal compressors are being replaced more and more by axial compressors, a series of rotating disks with large numbers of blades, with stationary bladed disks in between. The design of both compressors and turbines involves new aerodynamic problems, which lie in the field called the aerodynamics of internal flow, as distinguished from the aerodynamics of external flow, involved in the design of wings, fuselages, tail and control surfaces, and the like. The flow of compressible and incompressible fluids through a sequence of blade sections, called a *wing cascade*, is one of the basic problems of this new branch of aerodynamics.

Turbojets have the advantages of lighter weight and smaller frontal area than conventional engines. Their fuel consumption for the same performance is less favorable. Weight and fuel consumption are usually referred to unit thrust (pounds of fuel per hour and per pound of thrust) instead of unit power (pounds of fuel per hour and per horsepower). Turbojets with axial compressors are usually superior to those using centrifugal compressors, having smaller frontal area and smaller internal aerodynamic losses.

If we imagine that an airplane is flying very fast, say over 400 miles per hour, then the air that enters the engine can produce compression without any auxiliary device. This is called the *ram* effect. By use of the ram effect, we can simplify the engine by throwing away the compressor and turbine. The resulting device is called a *ramjet* (Fig. 69). It was proposed as early as

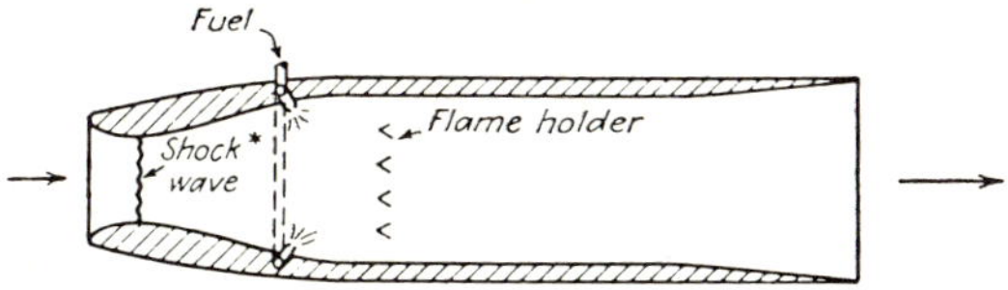

Fig. 69. Schematic diagram showing the elements of a ramjet. (From Joseph Liston, *Power Plants for Aircraft* [copyright 1953, McGraw-Hill Book Co., Inc.], by permission.)

1909 by René Lorin (Ref. 15). It possesses extreme mechanical simplicity but is penalized in comparison to the turbojet by higher fuel consumption—at least up to the flight-speed range of high supersonic Mach numbers—and by the fact that without a specific starting device it functions only above a certain flight velocity.

A very ingenious device that functions right from zero flight speed is the *pulse-jet* (Fig. 70). Like the ramjet it works without

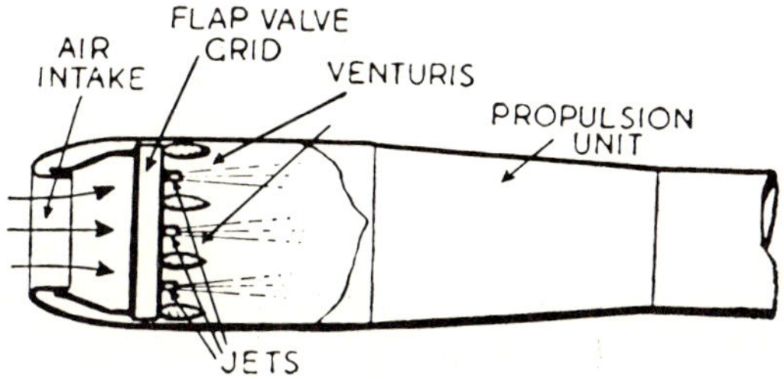

Fig. 70. Schematic diagram showing the elements of a pulse-jet. (Courtesy of *Flight*, London.)

compression and therefore does not need a turbine for compressor drive. It differs from the ramjet in that the process is not continuous but periodic. This device has intake valves which open and close somewhat as in a reciprocating engine, but the valves are controlled automatically, principally by resonance with the periodic process of successive compression, combustion, and outflow. The idea of such a device goes back to rather early years. The first practical application was made by the Germans; the device was known as the *Schmidt-Rohr* and was used for the propulsion of the so-called V–1 weapon, also called the buzz-bomb. The pulse-jet is well suited for target planes as an expendable propulsion device because of its low cost of manufacture, in contrast to the turbojet, which is a high-priced device. The manufacture of expendable turbojets has been suggested several times, but as far as I know has never yet been realized. New developments in pulse-jets seem likely to succeed in eliminating the valves and establishing the periodic process on the pure resonance principle, by appropriate choice of the relative dimensions of the components of the device (Ref. 16).

The relatively large manufacturing costs of turbojets and the relatively large consumptions of the ramjet and pulse-jet constitute a challenge to inventors to find devices more economical in thermal efficiency than the latter and cheaper than the turbojet. A class of such possible devices are the *wave machines*, in which the compression necessary for good thermal efficiency is produced by shock-wave action. They are, however, still in the stage of invention or, at best, in early development.

We proceed now to a short discussion of rockets, especially those using chemical propellants. We distinguish between rockets using *solid propellants* (Fig. 71) and those using *liquid propellants* (Fig. 72). The solid propellant is usually a mixture of an oxidizer and a combustible. It differs from an explosive like that used, for example, in bombs, in that it has a relatively slow burning rate. The burning may proceed through the thrust chamber in the axial direction (so-called cigarette burning, as in Fig. 71) or in the radial direction, both from inside and outside, as in many rockets used as weapons and containing "grains" in the shape of hollow cylinders. Finally, we have rockets with purely internal

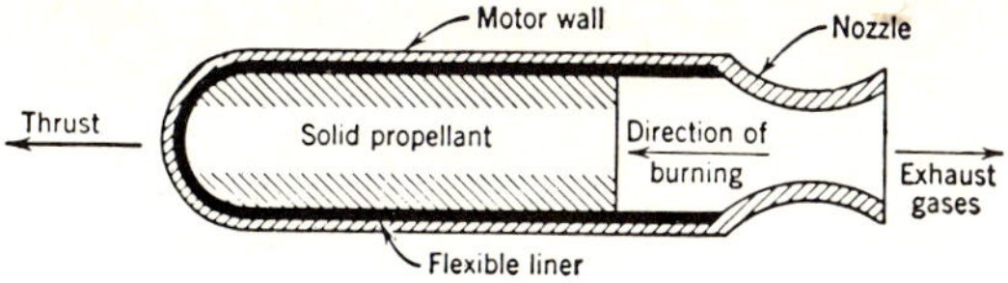

Fig. 71. Schematic diagram showing the elements of a solid-propellant rocket. (From M. J. Zucrow, *Jet Propulsion and Gas Turbines* [copyright 1948, John Wiley and Sons, Inc.], by permission.)

burning; i.e., the combustion proceeds from an inner hole. The art of grain design consists of devising the shape of the grain in such a way that a desired pressure-time relation is achieved. The total burning time may vary from half a second to 45 seconds according to the application: rockets for jet-assisted take-off of airplanes (JATO), boosters for missiles, forward-firing rockets as destructive weapons, and the like.

The liquid propellants can be classified as *monopropellants* and *bipropellants*. Monopropellants—e.g., a composition called nitromethane—generally produce oxygen and fuel by decomposition, and the result is the production of a high-pressure, high-temperature gas mixture. Another known monopropellant is ethylene oxide. With bipropellants, the fuel and the oxidizer are introduced into the thrust chamber separately (Fig. 72). Commonly used oxidizers are liquid oxygen, nitric acid, mixed nitric oxides, and fluorine. Common fuels include, for example, aniline, hydrocarbons, hydrazine, and ammonia. If a fuel-oxidizer combination is self-igniting, it is called a hypergolic combination. Hydrogen peroxide acts to some extent as a monopropellant since its decomposition, initiated by an appropriate catalyzer, produces considerable heat, and therefore it may be used alone in a rocket; however, one can complete the process by using up the surplus oxygen for burning an additional fuel. In this respect it acts as a component of a bipropellant system.

Solid propellants are stored in grain form in the thrust chamber, while liquid propellants are introduced through injectors, either

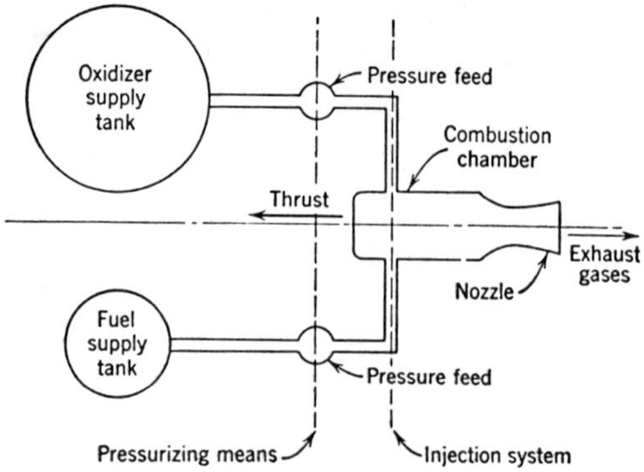

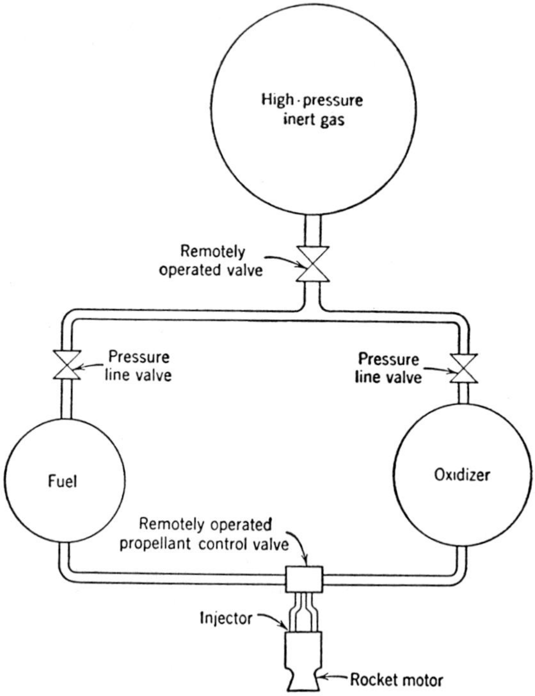

Fig. 72. Schematic diagrams showing the elements of two liquid-propellant rocket systems. The upper system uses pressurizing pumps in the oxidizer and fuel lines, while the lower employs a high-pressure inert gas to pressurize the propellants. (From M. J. Zucrow, *Jet Propulsion and Gas Turbines* [copyright 1948, John Wiley and Sons, Inc.], by permission.)

by means of pumps or from tanks put under pressure. In all rockets the intention of the designer is to obtain stable burning at constant chamber pressure. The jet leaves the combustion chamber through a nozzle, usually with a speed slightly above the velocity of sound corresponding to the high temperature of the combustion gases. For this purpose the cross section of the nozzle, after contraction to a smaller sectional area, is slightly expanded. The process of outflow is similar in all jet devices. It can be shown that if a compressible fluid leaves a chamber, which is at sufficiently high pressure, through a simply convergent nozzle, the velocity of the fluid at the exit is equal to the sound velocity corresponding to the temperature prevailing there. Some engineers concluded that a stream of gas or, in general, a compressible fluid cannot reach by expansion a velocity higher than sound velocity. The Swedish engineer, Carl Gustaf Patrik de Laval (1845–1913),[1] discovered that, if one wants to increase the velocity of gas in a nozzle beyond the velocity of sound, the nozzle, after converging to a minimum cross-sectional area, has to be expanded to a larger cross-sectional area. The velocity in the minimum cross section is—at least approximately (i.e., neglecting friction effects)—equal to sound velocity. Laval's principle of nozzle design is widely employed in turbines and jet engines.

In principle, nuclear reactors can be used in almost all jet-propulsion devices. We can imagine that nuclear reactors may replace the combustion chamber in a gas turbine or ramjet or the boiler in a steam turbine. The task of the reactor in this case is to introduce heat into air or water vapor. The main problem is to find methods to take heat out of the reactor and transfer it to air or vapor at sufficiently high temperatures; otherwise the efficiency is poor and the device becomes bulky. This involves technological problems of great difficulty. For manned airplanes the question of shielding, i.e., the protection of crew or passengers from radiation effects, is critical. Materials also have

[1] Also inventor of the cream separator.

to be protected against radioactive corrosion. To produce a nuclear rocket one might think of using jets of fission products directly for propulsion. A "photon" rocket has also been proposed. In such a device no mass is expelled from the rocket. The radiation pressure is directed to produce thrust. For the time being it seems more promising to use a working fluid, possibly of low molecular weight, so that high velocities can be realized at moderately high temperatures. The next decade probably will show what can be accomplished by the use of nuclear reaction in the field of jet propulsion.

Choice between Propulsive Systems

The general impression is that, except in small and medium-size transport and training planes, reciprocating engines are on their way out in the field of airplane propulsion. Helicopters, which receive increasing attention for moderate ranges in aerial transportation, both commercial and military, may prolong the lifetime of reciprocating engines. In long-distance aerial transportation the propeller driven by compound engines (combined piston engine and turbine) or by gas turbines (turboprop) is likely to prevail, at least for a few years, because of the superior propulsive efficiency of the propeller compared to the pure jet, certainly at high subsonic speeds.

The question of the appropriate choice of the best propulsive system for a given type of aircraft is a rather complicated and much-discussed problem. The first question is the comparison between the power required and the power available. The main goal of early calculations on the possibility of powered flight (see Chapter I) was to show that it is possible to make as much power available as the minimum power required for flight. In one of his talks dealing with the early period of aviation, Igor Sikorsky jokingly said that he had built an airplane which was unique in that its minimum, cruising, and maximum speeds were exactly equal! Actually the surplus of power available over the minimum power required essentially determines the performance of the aircraft.

The performance calculation for an airplane is the process of estimating its minimum and maximum speeds, its rate of climb as a function of the altitude, and its maximum range as a function of the assumed pay load. The cruising speed and cruising altitude are then determined from economic considerations. The performance calculation is similar for commercial and military planes, with the exception that for the military plane economic considerations may be secondary; the primary consideration is carrying out successfully the aircraft's mission.

A vertical riser is a plane which has sufficient propulsive power to lift itself along a vertical flight path. In the commercial field many people are now trying to create a convertible airplane, which may start either as a helicopter or a road vehicle and gradually go over to level flight.

Personally I have great confidence in the future of jet planes in the commercial field, although their higher fuel consumption and certain practical drawbacks such as, for example, excessive noise are still difficulties to be overcome. There is little doubt about the future of jet planes in military aviation; one of the important questions is how far ramjets and rockets will be used either as auxiliary or main means of propulsion. There are attempts—for example in France—to develop the ramjet as the main propulsive device for manned supersonic airplanes, using auxiliary turbojet or rocket devices to launch the aircraft to a sufficiently high speed for the ramjet to take over. The majority opinion at present, however, seems to be that the proper field for the ramjet is the unmanned missile.

The two main virtues of the rocket are concentration of very large power in a device relatively small and light and independence of atmospheric air. For these we pay a penalty in fuel consumption. In rocket engineering we mostly speak of specific impulse, defined as the product of thrust and its duration in seconds per unit weight of propellant. It is easy to understand that if we compute the specific impulse for a turbojet engine without considering the air as "propellant" (as against the rocket propellant, which contains its own oxidizer), we arrive at a number several times higher than that of the most efficient rocket. For example, the specific impulse of a turbojet that consumes 1 pound of fuel per hour per pound of thrust amounts to 3,600 seconds, whereas the usual value for liquid-propellant rockets is about 200 to 250 seconds.

Thus the rocket has two main fields of application: first, short-time production of large thrust, whenever manned airplanes or missiles need it, and, second, flight at altitudes where sufficient oxygen is not available. Rockets for assisted take-off and super-performance of manned aircraft and for boosters for the launching of missiles are in wide use. The German V–2 weapon had rocket propulsion exclusively, and similar weapon systems on the borderline between ballistics and aviation are in development in several countries. Finally, space travel by rocket ships is a popular subject both for science fiction and serious scientific research.

The rocket itself was probably a Chinese invention which started as a fire arrow. First bow and arrow were used to transport incendiary material, then the reaction of the combustion gases was used for propulsion of the arrow. However, an unconfirmed story also coming from China shows that rocket propulsion for flight was considered as early as around 1500. This story tells of an inventor named Wan-Hoo, who built a chair on two wheels and sat in the chair holding two kites in his hands for sustained flight. For take-off he attached forty-seven black powder rockets to his chair. According to the story, he succeeded in firing the rockets. After that, however, smoke and fire developed and Wan-Hoo, chair, and kites disappeared!

Through the centuries rocketry has been prominently used in parallel with gunnery. However, as the accuracy of gunnery was improved by the introduction of rifled guns, rockets temporarily lost their importance. The British General Staff decided, in the second half of the last century, that rockets no longer had military significance. Pyrotechnics, of course, continued to employ rockets. Signal rockets, also, have long been used, for example, in rescue work along the seacoast. Rocket developments for military purposes had a new boom during the Second World War. In the meantime, a number of forward-looking individuals have maintained an interest in large rockets and have toyed with the idea of space travel.

Space Travel

Speculation upon space travel is practically as ancient as speculation upon powered flight in the atmosphere. Legend and fiction contain many more or less fantastic descriptions of travels to the moon, around the moon, or to another planet. Some writers on the history of science credit Cyrano de Bergerac (Ref. 17) with predicting jet propulsion as a means for space travel as early as 1648 or 1649, when he wrote his account of a voyage to the moon. At the end of the last century a German mathematics teacher, Kurt Lasswitz, wrote a widely read interplanetary novel (Ref. 18), which, according to the testimony of the author's son, first referred to a space station. This station, however, was not a satellite traveling around the earth; it was suspended between Mars and the earth at a point where the gravitational forces are balanced. Shortly afterward, in 1903, Konstantin E. Ziolkowski, a Russian mathematics teacher, described a streamlined, rocket-driven vehicle for space travel which used liquid oxygen and hydrogen as propellants (Ref. 19). He was perhaps the first man to base his project on sound principles. His proposal included gyroscopic control and a jet deflector for navigation in space.

In this work it is not possible for me even to refer to many of the most interesting and relatively serious publications on the problem. I shall, however, mention Robert H. Goddard (1882–1945), who in 1919 in this country studied methods of reaching extreme altitudes (Ref. 20), and Hermann Oberth, who in 1923 in Germany published a book on rockets for interplanetary spaces (Ref. 21). Oberth was able to inspire a group of younger men to work on rocket design; this group was instrumental in developing the V–2 rocket projectile during the last World War. It appears that Oberth had little opportunity to contribute directly to the design of this rocket, which certainly represented great progress in the direction of long-range, high-altitude rockets. The V–2 rocket still holds the altitude record for a single vehicle. The highest altitude (242 miles) has been reached by a two-step rocket combination, consisting of a V–2 and a WAC Corporal; the latter was designed by Frank J. Malina at the Jet Propulsion Laboratory, California Institute of Technology. Oberth's best-known successes were his book, mentioned above, and a film produced by Fritz Lang of U.F.A. in Berlin (1929) and entitled "The Girl in the Moon," on which he collaborated as scientific adviser.

Wernher von Braun was originally one of the group of young enthusiasts who were directly or indirectly inspired by Oberth. I am convinced he would also be a magnificent adviser to any Hollywood enterprise in the field of space travel. However, his merits as promoter and organizer of the V–2 project (under the direction of General Walter Dornberger) and as a promoter of space-travel ideas in this country must be recognized.

"What are we waiting for?" says von Braun. "It will cost only five billion dollars! There are no problems involved to which we don't have the answers—or the ability to find them—right now." The layman is amazed and the expert is left wondering. I do not want to be either too skeptical or too enthusiastic.

Performance calculations for the vertical flight of a rocket escaping from the earth, and for a rocket serving as satellite around the earth, were carried out by several authors. The "escape" velocity U_e, i.e., the velocity necessary to escape from the gravitational force of the earth, is roughly estimated by the simple equation $\frac{1}{2}U_e^2 = gR$, which equates the kinetic energy of the unit mass to the work necessary to move the unit mass from the distance R to infinity against the gravity force, neglecting all other resistance. Substituting for g the value of the acceleration of gravity at the earth's surface and for R the radius of the globe, U_e becomes of the order of 7 miles per second.

The satellite velocity is somewhat dependent on the altitude at which the satellite is supposed to cruise. If the vehicle travels on a circular path of radius $R + h$, its velocity, U_s, must be great enough so that the centrifugal force will be equal to the acceleration of gravity at that altitude. If we take $h \cong 1{,}000$ miles, the satellite velocity is about 5 miles per second, and the circular orbit will be traversed in about 2 hours. Satellite trajectories leading from initial vertical flight to a circular path also were calculated by various authors.

The fundamental question is whether we can produce a rocket which would reach these tremendous velocities. Malina and Summerfield (Ref. 22), on the basis of an extrapolation from the data of rocket technology in 1946, calculated the ratio of the required weight of propellant to the initial total weight of a single rocket for reaching escape velocity. They obtained the following values corresponding to various propellant combinations:

Anilin + nitric acid	.995
Oxygen + alcohol	.991
Liquid oxygen + liquid hydrogen	.960

These figures mean that even in the case of the best propellant combination only 4 percent of the initial weight remains available for structure and pay load. It is evident that, except perhaps by the use of nuclear power, a single rocket has no chance of leaving the gravitational field of the earth. There remains the possibility of the "step-rocket," i.e., an arrangement in which parts of the structure, after the propellants carried in them are consumed, are left behind and only the last remaining "step" carries out the mission of the space vehicle. The step-rocket for space travel was suggested by Ziolkowski. Before him it was proposed by a French doctor by the name of André Bing (1911) for high-altitude research purposes. The concept appears to be much older; it is mentioned in the *Encyclopédie* of Diderot and D'Alembert.

A great number of authors have carried out more-or-less detailed and more-or-less reliable calculations on possible mass ratios. The reader may peruse, for example, the book by Willy Ley on *Rockets, Missiles and Space Travel* (Ref. 23), which contains much interesting historical and technical information. The result of these calculations is briefly the following: In order to reach escape velocity, it is hard to do better than to start with a rocket combination consisting of three step-rockets, where the initial take-off weight is about sixty-four times larger than the end product that would sail into the wide empty spaces of the universe.

Well, even if the initial weight of a man-carrying space rocket seems enormous at first thought, one can adjust one's mind to larger and larger figures. Let us, however, listen to a critic, in the person of Milton W. Rosen (Ref. 24), in charge of one of the United States Navy's important missile projects:

Altitude is the primary factor in any consideration of the feasibility of a manned, earth-returnable rocket. . . .

According to recent unofficial, but reliable reports, a Douglas Skyrocket [a rocket-powered airplane] has reached an altitude of 15 miles and its pilot has returned safely to the earth's surface. For an altitude of 15 miles, then, feasibility has been demonstrated. No more needs to be said.

Feasibility can be shown for a manned earth-returnable rocket that reaches an altitude of between 15 and 50 miles, even though no human has ever reached these heights. Rockets have been built which can ascend 50 miles and which can carry the necessary payload involved in transporting a human being. The significance of the 50-mile height is that parachute recovery has been successful below this altitude. Entire WAC Corporal rockets and instrument sections from Aerobees have been recovered by parachute from altitudes up to 50 miles.[2] Moreover, the accelerations encountered on the powered ascent are within the

[2] Mr. Rosen might have mentioned that one Aerobee rocket carried three monkeys and two mice in its instrument section to an altitude of 36 miles, in 1952. The United States Air Force succeeded in returning these personnel of the animal kingdom to earth by parachute without damage. By means of a motion-picture record, the behavior of the animals in gravity-free flight has been studied (U.S.A.F. Film No. 19832, "Animals in Rocket Flight," Aeromedical Laboratory, Wright Air Development Center).

tolerance limit of human beings. The maximum velocity is sufficiently low that, for a vertical ascent, the vehicle skin temperature will not tax the capacity of known materials and techniques of construction. The most important feature of a flight to less than 50 miles is that the duration of flight will be brief—a matter of several minutes. For this reason, many of the difficult problems that would be involved in flights to higher altitudes will be ignored when the altitude limit is only 50 miles. These problems include the effects on the vehicle and its passenger of cosmic and solar radiation, meteor collisions, and free fall in a vacuum. . . .

Above 50 miles the situation is entirely different. Attempts at parachute recovery of instruments have not been successful. Depending upon the altitude to be reached, the accelerations could be beyond human tolerance limits and vehicle skin temperatures above the melting points of available materials. The Viking rocket, which reached an altitude of 136 miles, could have carried a man, but no one could have insured his safe return. Moreover, no one could have calculated the probability of his survival; there are too many "unknowns." For example, if the duration of the flight is long, the effects of cosmic and solar radiations must be considered, but the nature and quantity of these radiations in outer space have not been fully determined and we are only beginning to study their effects on living cells. Another risk hard to assess is the danger of meteor collisions; although this hazard has been estimated and various schemes proposed for eliminating it, none have ever been tested. It is not possible to predict the physiological and psychological effects on a human being of weightlessness; a normal condition in space flight, but one rarely encountered in our earth-bound lives.[3]

One basic problem has to be mentioned—the problem of safe return to the earth or landing on any celestial body. Any rocket returning from space travel enters the atmosphere with tremendous speed. At such speeds, probably even in the thinnest air, the surface would be heated beyond the temperature endurable by any known material. This problem of the temperature barrier

[3] Reproduced by permission of the author and the American Technion Society.

is much more formidable than the problem of the sonic barrier. Even if it might be possible to arrange a gradual entrance into the atmosphere, approaching the earth by exact control along a spiral trajectory, it is improbable that a return can be achieved without using rocket power as brake. This, of course, means an enormous amount of fuel reserve. Unfortunately, we cannot imitate Lucian of Samosata (second century A.D.) who made his space-traveler hero, Menippus, return to earth in a very simple way: the God Mercury took hold of his right ear and deposited him on the ground.

The medical or biological problem of prolonged existence in a gravity field of practically zero intensity—a weightless existence—may be serious, and research in this direction appears highly desirable. Medical people may sometimes be too cautious. In *The History of Aeronautics* by Vivian and Marsh (Ref. 25) we learn that, at the time of the Montgolfier balloons, doctors were worried about altitude effects, since the general opinion was that the atmosphere does not extend beyond four or five miles above the earth's surface. So one day in 1783 they put a cock, a sheep, and a duck aloft as passengers in a balloon for an eight-minute ascent and descent. The duck and the sheep came through all right, but the cock was apparently affected by the rarefaction of the atmosphere. However, it came out later that the sheep had trampled on the cock, causing more physical injury than any that might be inflicted by rarefied air!

Let us return to the question of performance. It appears to me that the use of nuclear energy will make the rocket so much more efficient that serious attempts to build a space ship should await the advent of the nuclear rocket. For a rocket using hydrogen as working fluid and a nuclear reactor as heat source, the specific impulse of the working fluid can easily be made several times the present values for the usual propellants, without substantially raising the temperatures to which the rocket walls have to be exposed. Further developments in the use of nuclear processes for propulsion may allow even more significant improvements.

In the meantime, basic studies in aerodynamics and the physics of rarefied, ionized gases, gradual exploration of the highest altitudes reachable by sounding rockets, study of radiation effects on material and humans, study of navigation and guidance problems at high altitudes and in space, and development of unmanned rockets leading gradually to a satellite should give the enthusiasts of space travel enough to do. I do not believe in reckless promotion. On the other hand, I think that the "respectable" scientific and engineering societies should not close their doors to the astronauts or the pages of their journals to papers dealing with the problems of space travel. The present congresses and meetings of the astronautical and interplanetary societies have a relatively high scientific level, especially if we compare them with the activities of certain aeronautical societies only twenty-five years before the first mechanical flight.

Perhaps the effort necessary to proceed from the present-day long-range, high-altitude rocket to a manned space rocket is no more than the effort which led from the Wright brothers' airplane of 1903 to today's supersonic aircraft. This progress was achieved through the thinking and striving of two generations of practical engineers and theoretical scientists. I am satisfied if I have succeeded in presenting some of their problems, in a sketchy way, in the six chapters of this book.

References

1. Gabrielli, G., and Kármán, Th. von, "What Price Speed?" *Mechanical Engineering*, *72* (1950), 775–781.
2. Rankine, W. J. M., "On the Mechanical Principles of the Action of Propellers," *Transactions of the Institute of Naval Architects*, *6* (1865), 13–30.
3. Froude, W., "On the Elementary Relation between Pitch Slip, and Propulsive Efficiency," *Transactions of the Institute of Naval Architects*, *19* (1878), 47–57.
4. Drzewiecki, S., "Sur une méthode pour la détermination des éléments mécaniques des propulseurs hélicoidaux," *Comptes rendus de*

l'Académie des Sciences, Paris, 114 (1892), 820–822; *Des hélices aériennes; Théorie générale des propulseurs hélicoidaux, et méthode de calcul de ces propulseurs pour l'air* (Paris, 1909); *Théorie générale de l'hélice* (Paris, 1920).

5. Prandtl, L., Appendix to the paper of Betz mentioned in Ref. 6, *Göttinger Nachrichten, mathematisch-physikalische Klasse* (1919), 213–217.
6. Betz, A., "Schraubenpropeller mit geringstem Energieverlust," *Göttinger Nachrichten, mathematisch-physikalische Klasse* (1919), 193–213; reprinted by L. Prandtl and A. Betz in *Vier Abhandlungen zur Hydrodynamik und Aerodynamik* (Göttingen, 1927).
7. Helmbold, H. B., "Zur Aerodynamik der Treibschraube," *Zeitschrift für Flugtechnik und Motorluftschiffahrt, 15* (1924), 150–153, 170–173.
8. Wood, R. McK., and Glauert, H., "Preliminary Investigation of Multiplane Interference Applied to Propeller Theory," *Aeronautical Research Committee Reports and Memoranda* No. 620 (1918); Glauert, H., "An Aerodynamic Theory of the Airscrew," *Aeronautical Research Committee Reports and Memoranda* No. 786 (1922); "Note on the Vortex Theory of Airscrews," *ibid.* No. 869 (1922).
"Note on the Vortex Theory of Airscrews," *ibid.* No. 869 (1922).
9. Pistolesi, E., "Nuova sviluppo del metodo di Drzewiecki per il calcolo analitico delle eliche," *L'Aerotecnica, 3,* (1920), 147–173; "Nuovo indirizze e sviluppi della teoria delle eliche," *Atti dell' Associazione Italiana di Aerotecnica, 2* (1922), 28–44; "I piu recenti progressi nello studio dei propulsori elicoidali," *ibid., 3* (1923), 189–197.
10. Bienen, Th., and Kármán, Th. von, "Zur Theorie der Luftschrauben," *Zeitschrift des Vereines Deutscher Ingenieure, 68* (1924), 1237–1242, 1315–1318.
11. Goldstein, S., "On the Vortex Theory of Screw Propellers," *Proceedings of the Royal Society of London,* series A, *123* (1929), 440–465.
12. Moriya, T., "On the Induced Velocity and Characteristics of a Propeller," *Journal of Faculty of Engineering, Tokyo Imperial University, 20* (1933), 147–162.
13. Kawada, S., "Calculation of Induced Velocity by Helical Vortices and Its Application to Propeller Theory," *Report of Aeronautical Research Institute, Tokyo Imperial University* No. 172 (1939).

14. Schlaifer, R., "Development of Aircraft Engines," in R. Schlaifer and S. D. Heron, *Development of Aircraft Engines and Fuels* (Boston, 1950).

15. Lorin, R., "La propulsion à grande vitesse des véhicules aériens: Étude d'un propulseur à réaction directe," *L'Aérophile*, *17* (1909), 463–465; "Propulsion par réaction directe et son application à l'aviation," *ibid.*, *18* (1910), 322–325.

16. Logan, J. G., Jr., "Summary Report on Valveless Pulsejet Investigation," *Technical Memorandum* No. C.A.L.–42, Project Squid, Cornell Laboratory, Buffalo, N. Y., October, 1951.

17. Cyrano de Bergerac, Savinien de, *Histoire comique des états et empires de la lune* (Paris, 1656); *Histoire comique des états et empires du soleil* (Paris, 1662).

18. Lasswitz, K., *Auf zwei Planeten* (Leipzig, 1897).

19. Ziolkowski, K. E., *Rocket into Cosmic Space* (in Russian) (Kaluga, 1924). A reprint of an article which originally appeared in *Nautschnoje obozrenije* (*Science Survey*) in 1903.

20. Goddard, R. H., "A Method of Reaching Extreme Altitude," *Smithsonian Miscellaneous Collections*, *71* (1919), No. 2, 1–69.

21. Oberth, H., *Die Rakete zu den Planetenräumen* (Berlin, 1923); *Wege zur Raumschiffahrt* (Berlin, 1929).

22. Malina, F. J., and Summerfield, M., "The Problem of Escape from the Earth by Rocket," *Journal of the Aeronautical Sciences*, *14* (1947), 471–480.

23. Ley, W., *Rockets, Missiles, and Space Travel* (London, 1952).

24. Rosen, M. W., "The Prospects for Space Flight," *Technion Year Book* (1952–1953), 88–93.

25. Vivian, E. C. H., and Marsh, W. L., *The History of Aeronautics* (London and New York, 1921).

» *The Messenger Lectures*

IN its original form this book consisted of six lectures delivered at Cornell University in March, 1953, namely, the Messenger Lectures on the Evolution of Civilization. That series was founded and its title prescribed by Hiram J. Messenger, B.Litt., Ph.D., of Hartford, Connecticut, who directed in his will that a portion of his estate be given to Cornell University and used to provide annually a "course or courses of lectures on the evolution of civilization, for the special purpose of raising the moral standard of our political, business, and social life." The lectureship was established in 1923.

» *Index*